SYMBOLIC COMPUTATION

Artificial Intelligence

Managing Editor: D.W. Loveland
Editors: S. Amarel   A. Biermann   L. Bolc
        A. Bundy   H. Gallaire   P. Hayes
        A. Joshi   D. Lenat   A. Mackworth
        E. Sandewall   J. Siekmann   W. Wahlster

Springer Series
SYMBOLIC COMPUTATION—*Artificial Intelligence*

N.J. Nilsson: Principles of Artificial Intelligence. XV, 476 pages, 139 figs., 1982

J.H. Siekmann, G. Wrightson (Eds.): Automation of Reasoning 1. Classical Papers on Computational Logic 1957–1966. XXII, 525 pages, 1983.

J.H. Siekmann, G. Wrightson (Eds.): Automation of Reasoning 2. Classical Papers on Computational Logic 1967–1970. XXII, 638 pages, 1983.

L. Bolc (Ed.): The Design of Interpreters, Compilers, and Editors for Augmented Transition Networks. XI, 214 pages, 72 figs., 1983.

R.S. Michalski, J.G. Carbonell, T.M. Mitchell (Eds.): Machine Learning. An Artificial Intelligence Approach XI, 572 pages, 1984.

L. Bolc (Ed.): Natural Language Communication with Pictorial Information Systems. VII, 327 pages, 67 figs., 1984.

J.W. Lloyd: Foundations of Logic Programming. X, 124 pages, 1984.

A. Bundy (Ed.): Catalogue of Artificial Intelligence Tools. XXV, 150 pages, 1984. Second, revised edition, IV, 168 pages, 1986.

M.M. Botvinnik: Computers in Chess. Solving Inexact Search Problems. With contributions by A.I. Reznitsky, B.M. Stilman, M.A. Tsfasman, A.D. Yudin. Translated from the Russian by A.A. Brown. XIV, 158 pages, 48 figs., 1984.

C. Blume, W. Jakob: Programming Languages for Industrial Robots. XIII, 376 pages, 145 figs., 1986.

N. Cercone, G. McCalla (Eds.): The Knowledge Frontier. Essays in the Representation of Knowledge. 552 pages, 93 figs., 1987.

G. Rayna: REDUCE. Software for Algebraic Computation. 344 pages, 1987.

Gerhard Rayna

REDUCE

Software for Algebraic Computation

With a Preface by
Anthony C. Hearn
The RAND Corporation

Springer-Verlag
New York Berlin Heidelberg
London Paris Tokyo

Gerhard Rayna
Computer Science and Electrical Engineering Department
Lehigh University
Bethlehem, Pennsylvania 18015
USA

Library of Congress Cataloging in Publication Data
Rayna, Gerhard.
  Reduce: software for algebraic computation.
  Includes bibliographical references and index.
  1. REDUCE (Computer program)   2. Algebra—Data
processing.   I. Title.
QA155.7.E4R39   1987   512'.028'553629   87-20535

© 1987 by Springer-Verlag New York Inc.
All rights reserved. This work may not be translated or copied in whole or in part without the written permission of the publisher (Springer-Verlag, 175 Fifth Avenue, New York, New York 10010, USA), except for brief excerpts in connection with reviews or scholarly analysis. Use in connection with any form of information storage and retrieval, electronic adaptation, computer software, or by similar or dissimilar methodology now known or hereafter developed is forbidden.
The use of general descriptive names, trade names, trademarks, etc. in this publication, even if the former are not especially identified, is not to be taken as a sign that such names, as understood by the Trade Marks and Merchandise Marks Act, may accordingly be used freely by anyone.

Text prepared by the author in camera-ready form, using SCRIBE on a DEC 20 computer with a TALARIS 1200 laser printer.
Printed and bound by R.R. Donnelley & Sons, Harrisonburg, Virginia.
Printed in the United States of America.

9 8 7 6 5 4 3 2 1

ISBN 0-387-96598-X Springer-Verlag New York Berlin Heidelberg
ISBN 3-540-96598-X Springer-Verlag Berlin Heidelberg New York

Table of Contents

Preface	1
Introduction	6
1. Overview	11
1.1. Variables, values, assignments	11
1.2. The basic operations	15
1.3. Giving commands, getting answers	18
1.4. A first look at CLEAR	21
1.5. WS, SAVEAS, INPUT n	21
1.6. Some built-in functions	22
1.7. NUM and DEN	24
1.8. Differentiation	26
1.8.1. Partial derivatives	29
1.9. Integration	29
1.10. FOR ... DO	32
1.11. FOR ... SUM	34
1.12. FOR ... PRODUCT	36
1.13. WHILE ... DO	36
1.14. Boolean expressions	37
1.15. REPEAT ... UNTIL	40
2. A Harder Look	42
2.1. The SUBSTITUTION function	42
2.2. ARRAY and OPERATOR	45
2.2.1. Arrays	45
2.2.2. Operators	50
2.3. Matrices	54
2.3.1. MAT	58
2.3.2. Printing matrices	59
2.3.3. Matrix expressions	60
2.3.4. Other matrix operations	62
2.3.5. A matrix example	63
2.4. The COEFF function	65
2.4.1. Multi-dimensional arrays	67
2.4.2. Simple variable destinations	67

2.5. FACTORIZE	68
2.5.1. Simple variable destinations	70
2.6. The SOLVE function	71
2.7. LET and CLEAR	77
2.7.1. A first look at LET	77
2.7.2. LET power = ...	79
2.7.3. MATCH power = ...	82
2.7.4. LET product = ...	82
2.7.5. LET sum = ...	83
2.7.6. LET operator = ...	85
2.7.7. CLEAR	87
2.7.8. FOR ALL X LET ... = ...	88
2.7.9. FOR ALL ... SUCH THAT ... LET ... = ...	94
2.7.10. LET DF(...) =	97
2.7.11. Overlapping rules	98
2.7.12. WEIGHT	101
2.7.13. Complex assignments	104
2.8. WRITE	104
2.9. Grouping	107
2.9.1. Grouped statements	108
2.9.2. Groups as expressions	109
2.10. IF ... THEN	113
2.10.1. IF ... THEN statements	113
2.10.2. IF ... THEN expressions	114
2.11. PART and setting a PART	117
2.11.1. PART	117
2.11.2. Setting a PART	120
3. Setting Modes and Options	**123**
3.1. EXP	123
3.2. GCD	126
3.3. LCM	130
3.4. MCD	131
3.5. RESUBS	133
3.6. ORDER	134
3.7. FACTOR command	136
3.8. FACTOR switch	138

3.9. DIV	141
3.10. RAT	143
3.11. ALLFAC	143
3.12. LIST	145
3.13. NERO	147
3.14. NAT, FORT	148
3.15. PRI	149
3.16. KORDER	149
3.17. Domain modes	153
3.17.1. FLOAT	154
3.17.2. BIGFLOAT and NUMVAL	159
3.17.3. RATIONAL	163
3.17.4. MODULAR	165
4. Procedures	**169**
4.1. Procedures without parameters or RETURN	170
4.2. Procedures with RETURN	174
4.3. Returning multiple values	178
4.4. Procedures with one parameter	182
4.5. Procedures with more than one parameter	185
4.6. Procedures with local variables	189
4.7. Interaction of procedures	192
4.8. Linkage questions	194
4.8.1. Formal parameters	194
4.8.1.1. The copy variable	194
4.8.1.2. The LET exception	196
4.8.1.3. CLEAR	198
4.8.1.4. The SUB exception	199
4.8.1.5. Arrays and the like	200
4.8.1.6. Advice	201
4.8.2. Local variables	202
4.8.3. The scopes of variables	202
4.8.4. Exit on error	203
4.9. Procedures with GO TO	205
4.10. LET rules as procedures	206

5. Case Studies — 210
  5.1. Find the variables — 210
    5.1.1. Find a given variable — 210
    5.1.2. Finding all variables — 215
  5.2. Dividing polynomials — 220
    5.2.1. Exact division — 220
    5.2.2. Divisions with remainder — 221
      5.2.2.1. The linear denominator case — 224
    5.2.3. A polynomial division procedure — 224
    5.2.4. The REMAINDER function — 232
  5.3. LCM, GCD, and the Euclidean Algorithm — 236
    5.3.1. Least common multiple — 236
    5.3.2. Greatest common divisor — 238
    5.3.3. The Euclidean Algorithm — 239
  5.4. Systems of linear equations — 241
    5.4.1. A procedure — 244
  5.5. Series approximations to quotients — 248
    5.5.1. Maclaurin expansion — 251
  5.6. Families of polynomials — 252
    5.6.1. The Tschebycheff polynomials — 252
    5.6.2. The Newton polynomials — 260
    5.6.3. Orthogonal polynomials — 265
  5.7. Rationalizing denominators — 268
    5.7.1. Automating the process — 270
    5.7.2. Algebraic numbers as denominators — 275
  5.8. A bug involving surds — 278
  5.9. Noncommuting symbols — 282
    5.9.1. Quaternions — 283
    5.9.2. Steenrod Squares — 286

6. Running REDUCE — 289
  6.1. The basics — 289
  6.2. IN from files — 292
    6.2.1. REDUCE.INI — 295
    6.2.2. PAUSE, CONT, and DEMO — 296
  6.3. Making corrections — 297
    6.3.1. Correcting as you type — 297

Table of Contents   ix

6.3.2. Correcting the previous statement	297
6.3.3. Editing earlier statements	301
6.3.4. Correcting syntax errors in files	302
6.3.5. Correcting files	303
6.3.6. Correcting procedures	304
6.4. INPUT, RETRY, CMD	305
6.5. OUT, FORT, LINELENGTH	306
6.5.1. OUT to files	306
6.5.2. FORT	308
6.5.3. LINELENGTH and FORTWIDTH!*	310
6.6. COMPILE, FASLOUT, LOAD	311
6.6.1. Fast-loading files/fast-running procedures	311
6.7. TIME, SHOWTIME	313
6.8. DEFINE	314
6.9. Tracing	317
6.10. Expression input	318
6.11. Lost in LISP?	323
Index	**325**

Preface

CONTRIBUTED BY
DR. ANTHONY C. HEARN
THE RAND CORPORATION, SANTA MONICA, CALIFORNIA

REDUCE is a computer program for algebraic computation that is in world-wide use by thousands of scientists, engineers, and mathematicians. Although it traces its beginnings to 1963, until recently it has only been available on main-frame computers because of its relatively large resource requirements. In 1980 I predicted (1) that by the mid-1980's it would be possible to obtain personal computers in the $10,000 - $20,000 range capable of running REDUCE. I am therefore delighted to see that machines of the power of the IBM PC can now run this system, even though these computers are more modestly priced than my 1980 vision of the personal algebra machine. In addition to the need for the more widespread access that personal computers can now provide, there has been a longstanding need for a textbook to help the beginning user become better acquainted with the system. I am therefore very glad that Dr. Rayna has undertaken to write such a book, just as the era of the REDUCE personal algebra machine is beginning.

In order to understand the nature of REDUCE, a little history is in order. In 1963 I met Dr. John McCarthy, the inventor of LISP. He suggested to me that the LISP language was well suited for automating the hand calculations I was making as an elementary particle theoretical physicist, and, after a few experiments, I was convinced that he was correct. I have been working in the algebraic computation area ever since. The first publication concerning this work appeared in the August 1966 issue of the Communications of the ACM (2). This paper talked about the specific applications of algebraic techniques to elementary particle physics computations. It quickly became obvious that the techniques I was developing were quite general, and in 1968 the first paper describing a general algebra system "REDUCE" was published (3). "REDUCE" is not an acronym, although I continue to spell it in capital letters. Its name was actually intended as a joke: algebra systems then, as now, tended to produce very large expressions for many problems, rather than reduce the results to a more managable form. "REDUCE" seemed to be the right name for such a system.

REDUCE 2 first appeared in 1970. The big breakthrough in this release was that the whole system was written in an ALGOL-like dialect, now called RLISP, rather than the rather awkward parenthesized notation of LISP in which the original REDUCE was written. By this time, the system was being distributed to other users, thus marking the beginnings of a user community.

Whereas REDUCE 2 was essentially the work of a single person, REDUCE 3, first distributed in 1983, included several significant new packages that were the work of others: in particular, packages for analytic integration, multivariate factorization, arbitrary precision real arithmetic, and equation solving. The number of people now enhancing the system is measured in the dozens, plus the hundreds of people who take the time to report problems or suggest improvements. My heartfelt thanks go to all these people. They are of course too numerous to mention individually. However, I would like to single out for special thanks James Davenport, John Fitch, Martin Griss, Steve Harrington, Fujio Kako, Jed Marti, Arthur Norman, Julian Padget, Tateaki Sasaki, Eberhard Schruefer and David Stoutemyer, for having made significant contributions to the design and development of the system.

A commitment to release updated and expanded versions of the system was made from the beginning, since I realized that the techniques of computer algebra would continue to develop and improve, thus requiring an algebra system to change to keep pace with these developments. We now try to release new versions of the system at yearly intervals, containing new capabilities, improved programming techniques, and bug fixes. (Unfortunately, current software practices do not enable anyone to produce bug-free a system with the range and complexity of REDUCE!)

From the beginning, REDUCE was designed with a number of definite goals in mind. One such design goal was portability. In the early days, the computing requirements were relatively high compared to the resources available. As a result, I had to be sure that I could use whatever computing equipment was made available to me as effectively as possible. Using LISP provided a certain level of portability, in that I could use any machine that had a LISP processor. However, as time went on, the LISP language itself began to evolve into different dialects. Consequently, the availability of a "LISP" on a given computer no longer guaranteed that I could use that machine. To compensate for this, I limited the REDUCE implementation to depend only upon a specific subset of LISP that one could find either directly or by simple mappings in all of the available LISP implementations. This led to Standard LISP (4), a uniform subset of LISP that could be easily implemented on any computer that already supported a working LISP system. Initially, we would map the Standard LISP subset onto the LISP of the target machine. However, as more programming tools were written in Standard LISP itself, culminating in a complete portable LISP compiler in 1981 (5), we chose instead to force the LISP on which we were running to conform to the Standard LISP definitions, so we were running Standard LISP itself rather than some other dialect.

One consequence of this activity has been that when implementors target a new machine for LISP development, they know they can run REDUCE if they can remain compatible with the Standard LISP protocols. In particular, they can use the portable compiler for producing high quality efficient code. A look at statistics we have collected from running REDUCE on a variety of different computing systems that use the portable compiler show a strong correlation between the published execution speeds of the machines and the times for running standard tests (6). Invariably, the times for other LISP implementation models are slower than the Standard LISP model.

The LISP standards we have adopted have enabled us to implement REDUCE on a wide variety of machine architectures with essentially no changes in the REDUCE source files themselves. Of course each machine requires some system dependent support to allow for differences in such things as input, output, and character sets. However, this support has been kept to a minimum. As a result, we have been able to implement the full REDUCE on at least twelve architectures ranging in power from the IBM PC to the Cray X/MP, a ratio of REDUCE execution speeds of over 500 to 1.

Another goal that has been very important in the development of REDUCE has been modularity. In order for a system as large and complicated as REDUCE to be maintained and extended, it is necessary that new facilities can be added without requiring changes to the existing code. In particular, a knowledgeable user should be able to add a new application with some assurance that his program can coexist happily with the existing code. To achieve this goal, many of the facilities in REDUCE 3 are written to depend only on the underlying Standard LISP subset, and interface to the rest of the system through entries in various system tables. In this manner, we were able to add facilities for handling special number systems such as arbitrary precision reals and modular arithmetic without requiring any major system changes.

From the start, REDUCE was designed to be used interactively. This is not unusual in these days of personal computers, but was not common when the system was first produced. As you will see from this book, many problems can be formulated by simply writing expressions, if necessary augmenting the built-in algebraic capabilities of the system by simple rules defining particular transformations needed. However, REDUCE also provides the user with a complete programming language when necessary. The language includes control structures such as FOR and WHILE, block structures, procedure definition, and a variety of algebraic and symbolic data types.

One of the most important attributes of REDUCE is its worldwide acceptance as a useful problem solving tool. As a result, there is a well-established base of knowledge about the use of the program in a wide variety of application areas. These areas include quantum electrodynamics and quantum chromodynamics, celestial mechanics, fluid mechanics, general relativity, numerical analysis, plasma physics, and a variety of engineering problems such as electrical network analysis and turbine and ship hull design. This maturity provides the user with some assurance about its reliability and ease of use.

OCTOBER 1986

REFERENCES

(1) Hearn, A.C., "The Personal Algebra Machine", Information Processing 80 (Proc. IFIP Congress 80), North-Holland, 621-628, 1980.

(2) Hearn, A.C., "Computation of Algebraic Properties of Elementary Particle Reactions Using a Digital Computer", Comm. of the ACM, 9, 573-577, 1966.

(3) Hearn, A.C., "REDUCE - A User Oriented Interactive System for Algebraic Simplification", Interactive Systems for Experimental Applied Mathematics, 79-90 (edited by M. Klerer and J. Reinfelds), Academic Press, New York, 1968.

(4) Marti, J.B., A.C. Hearn, M.L. Griss, C. Griss, "Standard LISP Report", SIGPLAN Notices, ACM, New York, 14, 48-68, 1979.

(5) Griss, M.L., A.C. Hearn, "A Portable LISP Compiler", Software Practice and Experience 11, 541-605, 1981.

(6) Marti, J.B., A.C. Hearn, "REDUCE as a LISP Benchmark", SIGSAM Bulletin, ACM, New York, 19, 8-16, 1985.

Introduction

The reader certainly doesn't need to be told that the use of computers for computing, for getting numerical answers to numerical problems, is widespread. So is the use of computers to manipulate symbolic information in the form of address lists, shopping lists, and so on -- usually referred to as word processing.

The use of computers for symbolic mathematics, such as ordinary Algebra and Calculus, is not as well known. Yet a number of programming systems making this possible are widely available. One of these, REDUCE 3[1] (hereafter referred to simply as REDUCE), is the subject of this book.

An algebraic computation system has many possible roles. It can be used to repeat or check simple hand algebra, making use of its infallibility more than its speed. It can be used to perform computations which are only slightly more complicated than ones the user would willingly carry out himself. And it can be asked to perform calculations orders of magnitude larger than reasonable for humans.

As a simple example, suppose one wants to apply the formula for rotation of axes to the equation of an ellipse. He can proceed as follows:

Tell REDUCE the rotation equations:

X := X1 * COS A + Y1 * SIN A $

Y := -X1 * SIN A + Y1 * COS A $

The ellipse is described by an equation in which a certain expression is set equal to zero. Type in that expression:

---

[1] Specifically, REDUCE 3.2, Copyright 1985 by The RAND Corporation, Santa Monica, California. All Rights Reserved.

Introduction

```
ELL := (X/H)**2 + (Y/K)**2 - 1 $
```

(The **2 means "squared".) Knowing that the answer will have a denominator which is unimportant, we ask REDUCE to print out just the numerator of the answer:

```
NUM ELL;
```

And here it is:

$$X1^2 * (COS(A)^2 * K^2 + SIN(A)^2 * H^2)$$
$$+ 2*X1*Y1*COS(A)*SIN(A)*(- H^2 + K^2)$$
$$+ Y1^2 * (COS(A)^2 * H^2 + SIN(A)^2 * K^2) - H^2 * K^2$$

As another example, suppose we have been given a formula which, it is claimed, has the value SIN T if the value of the variable U in it is TAN (T/2) -- and we don't quite believe it.

So we define that variable:

```
U:=TAN (T/2) $
```

And type in the formula we were given:

```
A := 2*U/(1+U**2) $
```

Now for some versions of REDUCE that's all we need, but some versions need to be "reminded" of some standard trigonometric identities:

```
FOR ALL X LET
    TAN X = SIN X / COS X,
    (COS X)**2 = 1 -(SIN X)**2,
    SIN X * COS X = SIN(2*X)/2;
```

So let's see what A is now:

```
A;
```

```
SIN(T)
```

So the formula was right! To evaluate A, REDUCE used all the relevant simplifying rules it knew about.

REDUCE is designed more for convenience than for the ultimate in efficiency. Its limit of usefulness falls somewhere between the categories

- computations that are only slightly more complicated than ones the user would willingly carry out himself, and

- calculations orders of magnitude larger than reasonable for humans.

Let us use the term <u>superdifficult</u> for problems in the second category. For superdifficult problems a specially coded program, in extreme cases written in machine language, may be the most appropriate tool. REDUCE's strength lies instead in the ready availability, within it, of general-purpose algebraic simplification mechanisms. Superdifficult problems usually require more special techniques.

REDUCE can be used on several levels. First, it can be treated as a "hand calculator" which that can add, subtract, multiply, etc. not only numbers but algebraic expressions involving variables. On a more subtle level we can learn how to define procedures (and create whole libraries of procedures) for operating upon expressions. Deepest of all, we can learn how to add basic features to the REDUCE system (or modify existing basic features) by writing procedures in RLISP, the implementation language of REDUCE. This book will not attempt to explain the use of RLISP.

Even if we remain on the hand calculator level, there is much to study, since the user will want to learn how to add to the built-in environment of so-called LET rules which direct the automatic substitutions and simplifications during computation.

Introduction 9

While this book doesn't assume prior familiarity with any particular programming language, the reader who has never programmed a computer in a language like BASIC, Pascal, or FORTRAN may find it more difficult to get started with REDUCE. Readers who know Pascal (or ALGOL), in particular, will find many similarities, but should watch out for differences. We will occasionally point out places where Pascal programmers are likely to be led astray by inappropriate reliance on Pascal-derived habits. Readers who don't know Pascal should ignore those warnings!

A word about the exercises embedded in the text. They are an essential part of the exposition, and shouldn't be treated as if they were intended only as routine drill. The reader is asked to discover for himself, by doing the exercises, facts about the behavior of REDUCE which in many cases are not explicitly stated in the surrounding text.

The author tested all the examples in this book using REDUCE 3.2 on a Digital Equipment Corporation DEC 20 system, and most of them using REDUCE 3.2 on a Zenith 158 "IBM-compatible" personal computer. The illustrations should be valid on any computer with REDUCE 3.2, because the same software (written in RLISP) is used on all computers. (In turn, RLISP is supposed to behave in exactly the same way on all systems.) Exceptions may be encountered if, because of limited file or memory space, some features of REDUCE 3.2 had to be omitted from the system on some computer. Operating system dependent aspects, such as character sets, file naming conventions, and access -- if any -- to a standard text editor, may also vary.

REDUCE undergoes changes from time to time to add new features, improve its efficiency, and remove limitations. As a result, it may happen that in a future version some of our examples will no longer work as described here, because they depend on some aspect of REDUCE that came to be regarded as a "quirk" to be eliminated.

The author wishes to acknowledge the assistance of Dr. Anthony C. Hearn of The RAND Corporation, Santa Monica, California, the author of REDUCE, for supplying support and information, for careful reading of the manuscript, and for contributing the Preface to this book; of Dr. Jed Marti, also of The RAND Corporation, for additional proofreading and comments; and of Lehigh University for providing him with computer facilities for preparing the book and with students and faculty members on whom to try out earlier drafts.

1. Overview

A study of this Overview of REDUCE, together with the "Running REDUCE" chapter, should enable one to make quite profitable use of this algebraic manipulation system.

Topics deferred to Chapters 2, 3 and 4 include the use of arrays and matrices, the definition of procedures, and the use of files. Also postponed is discussion of the many optional modes that are available for modifying computational procedures, output forms, number systems used, and so on. The impatient reader with a particular problem in mind is welcome to jump ahead, especially to Chapter 2, for information he may need.

Chapter 5 is a collection of case studies: discussion of the solution of quite a number of types of problems. Some of these solutions required overcoming a succession of surprising complications. These are analysed in detail. The reader should work his way through at least this Overview, doing most if not all the exercises, before looking at that Chapter.

1.1. Variables, values, assignments

As in other computer languages, operations in REDUCE deal with variables and their values. But in contrast with more familiar computer languages, REDUCE variables have values which can not only be numbers but algebraic expressions involving other variables:

   1234,   1234/5679,   U+V,   3*ANGLE-(5*X**(3/4) + Y**10)

When we ask REDUCE to carry out an operation like A*B or ANGLE+STEP, we are asking it to combine the expressions which are the values of those variables (the values of A and B to be multiplied, the values of ANGLE and STEP to be added), collect terms, and -- if possible -- simplify the result.

The meanings of the symbols +, -, *, and / are obvious. The symbol ** means raised-to-the power. On some REDUCE systems it can also be written as ^.

Names of variables can be written in capital or lower case letters, or even mixed. REDUCE considers ANGLE, angle, and aNgLe to be the same variable. In this book we will use all capital letters; some people prefer all lower case.

Numbers in REDUCE are (normally) integers or quotients of integers: that is, whole numbers or fractions, not decimals. The number 13.579 would be represented as 13579/1000. This way answers are always exact if the original data can be given exactly -- there is no roundoff. There is also no limit (other than the size of the computer memory) on how many digits a number can have: the number 3**1000 (3 to the 1000th power), for example, can be handled effortlessly, even though it takes seven lines for REDUCE to display its value. (Among the many options in REDUCE is the use of other number systems, such as multiple precision floating point with any specified number of digits -- of use if roundoff can be tolerated in, say, the 100th digit.)

A variable is given a value by using the assignment symbol := , as in the examples

A:=1234;   APPLE:=RED*FRUIT;   WW:=(X+Y)**100;

If the variable already has a value, it can be changed in the same way, e.g.

APPLE:=APPLE+WORM;   WW:=(X-Y)**100;

The value of a variable can be as simple as a number, or as complicated as an expression involving numbers, other variables, and other forms we will learn about later.

The name of a variable can be of any length. It can be made up of letters and digits, but must begin with a letter. Other characters can be used, such as -, *, or even blank, but they must be preceded by an exclamation mark ! when typed:

THIS! IS! A! FUNNY!-NAMED!*VARIABLE

Variables, values, assignments

There are over a thousand identifiers (variable names and procedure names) which have some special significance in REDUCE. There is no reasonable way to memorize them or even to list them. But most of these names have the character * (typed !*) in them, so if the user avoids *-names the danger of collision with a "reserved word" is greatly reduced. However, E and I are reserved variables and can only be used -- except in special contexts -- to mean what they usually mean in mathematics: E to be the base of the natural logarithms, and I to be the square root of -1.

It is difficult to resist the urge to use the letter E as a variable, either in a sequence of variables A, B, C, D, E, ... , or by itself standing for Expression. In this book we often use the name EE for a variable where E would have been used were it not reserved. Likewise, the letter T can not generally be used as the name of a variable, either. It has a special meaning in the underlying language LISP.

As a general rule, if something you are doing in REDUCE produces very cryptic error messages, consider the possibility that you have inadvertantly used a reserved word, and change the names of your variables. For example, the response to the reasonable-looking statement

    SUM := SUM + X;

is an "Improper Delimiter" message! The trouble comes from SUM being a reserved word, used in the FOR ... SUM construction described later in this chapter.

Of great importance in REDUCE -- and nonexistent in most other computer languages -- is the concept of a variable being clear, i.e. not having an assigned value. For example, if no ABC:=... assignment has yet been made during the present run of REDUCE, then the variable ABC is said to be clear.

If a variable is clear, its value is itself (more technically, its own name): thus the value of ABC would be ABC. If a variable is not clear, its value is obtained by taking its assigned value (which is some expression), seeing if any of the variables in it have assigned values, substituting and simplifying these values, and repeating the process until a value is obtained made up only of numbers, clear variables, and other forms that can not be further simplified.

By the "assigned value" of a variable is meant, of course, the value the right-hand side of the := had the last time an assignment was made to the variable. The (current) value differs from this if the assigned value contains variables to which values have since been given.

To give an example, after the sequence of assignments

B:=Q;

U:=(X+Y+Z)**2;

Y:=10;

X:=A+B;

the value of U would be (A+Q+10+Z)**2, in view of the current values of X, Y, and B. (Note: under the standard options in REDUCE, the value of U would be printed out not as (A+Q+10+Z)**2, but expanded out and with the exponents (all of which happen to be 2) raised to the line above the rest of the formula.)

Exercise 1.1.1.

Assume W is not clear. Does the assignment W:=W ever accomplish anything? Explain. <u>Hint</u>: distinguish carefully between the (current) value of W and the assigned value of W, and consider the value of W after the steps

W := A;

A := 1;

W := W;

A := 2;

Variables, values, assignments 15

What would the value be if the W:=W were not there?

1.2. The basic operations

Expressions are built up from variables and numbers by the use of the operations +, -, *, /, and **, aided by parentheses (...). When more than one of these operations appears in an expression, the "usual" grouping rules of algebra apply. We review these here.

If only + and - signs appear, the additions and subtractions are assumed performed in order from left to right.

A+B+C-D+EE+F

means: Start with A, add first B and then C, subtract D, and finally add EE and then F to the result. This is so even if spaces are inserted (spaces are ignored, in fact):

A+B+C   -   D+EE+F

If the sum D+EE+F is to be subtracted, parentheses must be used:

A+B+C-(D+EE+F)

or more symmetrically

(A+B+C)-(D+EE+F)

Alternatively, we can change the last two plus signs to minus:

A+B+C-D-EE-F

Few readers are likely to make errors in the above. But for * and /, for which exactly the same principle applies in REDUCE, the habits formed in studying algebra in high school are not as adequate. We tend to interpret the algebraic expression $w/2a$ to mean w divided by the product $2a$. In REDUCE we must use a multiplication sign between the 2 and the A, and W/2*A is interpreted as $(W/2)$ times A, that is, as if the A were in the numerator. We can illustrate this in a manner exactly analogous to that for + and -:

A*B*C/D*EE*F

means: Start with A, multiply first by B and then by C, divide by D, and finally multiply the result first by EE and then by F. This is so even if spaces are inserted:

A*B*C   /   D*EE*F

If the product D*EE*F is to be interpreted as being in the denominator, parentheses must be used:

A*B*C/(D*EE*F)

or more symmetrically

(A*B*C)/(D*EE*F)

Alternatively, we can change the last two * signs to /:

A*B*C/D/EE/F

If the expression has both + (or -) signs, and * (or /) signs, and no parentheses, the multiplications and divisions are interpreted as being performed first, and the results then added and subtracted as indicated. For example,

A-B   /   A+B

does not, in spite of the deliberately misleading spacing, mean "the difference divided by the sum", but means "A, minus the quotient B/A, plus B". To represent the difference divided by the sum, parentheses must be used:

(A-B)/(A+B)

In fact, it's good practise when entering a fraction always to put the numerator and the denominator in parentheses, unless that part is a single variable or number.

As has surely been made clear already, parentheses can be used freely to enforce different groupings of operations, or just to make the grouping more evident. Only parentheses, "(" and ")", can be so used, not square brackets "[" and "]".

The basic operations   17

We note here that REDUCE allows + and - to be used as unary operations, that is, in front of the first term of an expression:

    Q1 :=   -A - B;

    Q2 :=   -A + B;

    Q3 :=   +A - B;

    Q4 :=   +A + B;

The + signs in front of the A in the last two lines have no actual effect, but can be written if desired to emphasize the absence of the - signs which appeared in the earlier lines.

If expressions beginning with unary minus signs are to be multiplied or divided, they should be enclosed in parentheses as a matter of good form, even though REDUCE actually allows expressions such as A/-B.

It remains to make two points about the grouping behavior of the exponentiation operator **. A**B**C is interpreted as A raised to the B power, and the result raised to the C power. This is mathematically equivalent to A**(B*C), A raised to the B*C power. If the other grouping is desired, parentheses must be used: (A**B)**C.

The ** operation applies only to the single number, variable, or parenthesized expression immediately in front of it. For example, in 2*A**5 only the A is raised to the power 5, not the 2. Similarly, -A**6 is -(A**6), since it's the A and and not the -A which is raised to a power. As still another example, COS(X)**3 means COS(X**3). If it is the cosine which is to be raised to a power, we must write (COS X)**3.

The single most important rule to remember is: when in doubt, use parentheses.

1.3. Giving commands, getting answers

The reader will have noticed that in each illustration of the assignment command we followed the command with a semicolon. This is because REDUCE is not a line-oriented language: it doesn't pay attention to the division of the stream of input into lines. It only pays attention to the so-called terminator characters: the semicolon and the dollar sign.

If a command is terminated with a semicolon (followed by striking the RETURN key), the command is obeyed and the result, if any, is printed. If a command is terminated by typing a dollar sign (followed by striking the RETURN key), the command is obeyed but the result is (normally) not printed.

The := assignment operation does have a "result", and, if the command ends with a ";", a report of the assignment is printed out, showing the current value of the right-hand side. For example, if U and V are clear, in response to the command

X:=2*U+3*V-U+7*V;

the "result" X := U + 10*V is printed out automatically.

Some other types of command, such as the CLEAR command described in the next section, do not have a "result", so nothing is printed by REDUCE in response even if a semicolon is used.

In examples printed in this text we indent the output produced by the computer. This doesn't actually happen when REDUCE is run. (This is not the only difference between the output as produced by REDUCE and the output as reproduced here. To make it fit it better on the printed page, the division of the output into separate lines has in many cases been changed.) Explanatory comments begin with a percent sign % (repeated on each line of any multi-line comment).

To illustrate,

Giving commands, getting answers

```
X:=2*U+3*V-U+7*V;          % we assume U, V clear here
        X  :=  U  +  10*V

U:=A+B;                    % we assume A, B clear here
        U  :=  A  +  B

V:=A-B;
        V  :=  A  -  B

X:=2*U+3*V-U+7*V;          % same input with U, V
                           %    no longer clear
        X  :=   11*A  -  9*B
```

Instead of entering complete assignments, we can type in just the name of a variable followed by a semicolon and RETURN, or an expression followed by a semicolon and RETURN. Then the current value of the variable or of the expression is determined and is automatically printed out. For example, if U and V were as above we would get the responses

```
U;
        A  +  B

U+V;
        2*A
```

Several commands separated by semicolons can be entered on one line, and one command can be broken into several lines eventually ended by its semicolon (and RETURN). If a line was typed without a terminating semicolon, and RETURN is struck, REDUCE waits for more input to be typed on the next line. When you realize that this is why nothing seems to be happening, just type the semicolon on the next line by itself, and strike RETURN again!

If a command doesn't fit on one line, and a printed record of the input is being maintained, divide the command at points chosen for maximum clarity and readability. Conversely, several short commands that "belong together" can be typed on the same line for the sake of better documentation. REDUCE is concerned only about the presence or absence of terminators.

In the examples we give in this text we will usually use the semicolon as terminator, unless we want to emphasize that the result is of no interest, or would take unreasonably long to print out. In the latter cases we will use the dollar sign, $.

We have already mentioned the fact that when REDUCE prints out expressions with exponents it uses the format generally found in algebra, with the powers raised, not the in-line form used for input:

A**2+B**2;

$$A^2 + B^2$$

If input-compatible output is needed, REDUCE can be instructed to change to that mode. See the OFF NAT command in Chapter 3.

Exercise 1.3.1.

Find out how to access REDUCE on a computer available to you. Then try all the examples given so far!

Exercise 1.3.2.

Use REDUCE to find the result of substituting Y-5 in place of X throughout the expression

$$X^3 + 15*X^2 + 10*X + 20 \quad .$$

Exercise 1.3.3.

With X clear, what would you expect X := X+1 to do? Try the following. <u>Caution</u>: don't try it unless you know how to stop (abort) misbehaving programs on your computer.

X := X+1;

X;

Would you expect the same result from X := X with X clear? At this time you only know one way to get X to be clear: stop this REDUCE session and start REDUCE all over again. Do so and try

Giving commands, getting answers 21

it.

If you want to continue using REDUCE after this second test, without again ending the session and restarting, first "repair" the variable X by assigning it a different value, say zero, by entering X := 0; .

1.4. A first look at CLEAR

We referred to a variable as being clear if no assignment has been made to it during the present REDUCE session. If an assignment has been made, REDUCE can be told to forget it, and return the variable to its original state, by giving the command

CLEAR ABC;

If several variables are to be cleared, you can use separate CLEAR commands, or combine them:

CLEAR ABC,PQR,LMN;

After such a CLEAR, the value of ABC would again be ABC; of PQR, PQR; etc.

1.5. WS, SAVEAS, INPUT n

If we generate a result which we hadn't thought of assigning to a variable, and realise we want to use it further, we can save it as a variable even after it's calculated and printed. We have a choice of two notations:

W := WS$

or

SAVEAS WS

WS is a reserved name meaning Workspace, and always refers to the result of the last "ordinary" statement executed. (It is left unchanged by the CLEAR command and by many of the statement types which we have yet to introduce, such as LET and ARRAY.) SAVEAS W is essentially identical in function to the first notation. It may be easier to type. The WS notation has the advantage that with it we don't have to save WS "as is": if, for example, we want to save the square of the last result, we can use

```
W := WS**2$
```

If at the same time we want to see what we saved, we can use a semicolon instead of the dollar sign.

In most versions of REDUCE, each input command during a session is prompted with a command number -- 1: for the first command, 2: for the second, and so on. The result, WS, of every command, say of command no. 73, is saved for the remainder of the session and is available by referring to WS 73 or, if preferred, WS(73).

In these same versions of REDUCE, all the actual commands typed during the session are also saved. Suppose that as command number 54 you typed a long expression involving the variable A. If you then change the value of A and want to recompute the expression, you don't need to retype the long expession: just enter the command INPUT 54; or INPUT(54); .

Of course for these features, WS n and INPUT n, to be useful it's necessary to know the number n of the result to be fetched or the command to be re-issued. The command DISPLAY 10$ lists the 10 most recent commands (in reverse order). If after the word DISPLAY something is typed which is not a number, all commands back to the beginning of the session are listed. (It's customary to use DISPLAY ALL$.)

1.6. Some built-in functions

REDUCE knows that SIN, COS, LOG, SQRT, etc. are names of functions, but knows very little more than that about them. The point is that you can write things like SIN(A+B) in expressions, and REDUCE will handle the entire symbol -- SIN(A+B) -- like a

Some built-in functions   23

variable. (It will, of course, first evaluate A+B if A and B aren't both clear.) It knows simple things such as SIN(0) = 0, and SQRT(144) = 12, but in the normal mode of operation it will leave forms like SIN(A+B) or SIN(11/7) unchanged. Such forms are referred to as "kernels", or, for emphasis, "unevaluated kernels".

Exercise 1.6.1.

Test to see if the version of REDUCE you are using knows the exact values of SIN(0), COS(0), SQRT(144), SQRT(A**2), (SQRT(B))**2, LOG(1), and similar expressions. Does it simplify SQRT(288) to 12*SQRT(2)? How about SQRT(-288)?

Note that REDUCE simplifies SQRT(A**2) to A, and not to -A. Yet if the value of A will eventually be a negative number, like -5, the simplified answer A is actually wrong, since $SQRT(25)$ is not A which is -5, but -A which is 5. This is our first encounter with the fact that REDUCE unavoidably doesn't always give the same answer that our usual mathematical conventions require. Another example of this is that REDUCE simplifies X/X to 1, which is "wrong" if X is 0 (in which case the answer should ideally be the word INDETERMINATE or something similar).

In a later section we will explain how you can use LET to specify more properties of these and other functions.

We will also explain how to activate a set of built-in procedures for finding numerical approximations (to any number of decimal places!) for many of these functions.

This is as good a place as any to mention that properties can be supplied, or are built in, not only for functions but also for variables. The symbol I is intended to be used to represent the square root of -1, so I**2 is (normally) automatically converted to -1. The symbol E has the property that LOG(E) is (normally) replaced by 1. And E and PI are replaced by their values, to any specified number of decimal places, if the modes BIGFLOAT and NUMVAL have been set ON. (See Chapter 3.)

Exercise 1.6.2.

Test to see if the version of REDUCE you are using knows the exact values of SIN(PI), SIN(PI/2), COS(2*PI), LOG(E), LOG(E**2), and similar expressions.

1.7. NUM and DEN

NUM and DEN are functions that take a single expression, simplify it, and return the numerator or denominator of the result.

```
NUM(X/Y**2);
                X

DEN(X/Y**2);
                2
               Y

NUM(100/6);
               50           % since 100/6=50/3

DEN(100/6);
                3

L:=(X*(X+3))/(X*(X+5))$

NUM (L);
              X + 3         % since an X cancels
```

In its normal mode of operation REDUCE only cancels those common factors between the numerator and the denominator that are "easy to find", as was the case for the common factor X in constructing L above. An example in which this fails follows:

NUM and DEN

```
A:=(X+1)*(X+2);
```

$$A := X^2 + 3*X + 2$$

```
B:=(X+1)*(X+3);
```

$$B := X^2 + 4*X + 3$$

```
C:=A/B;
```

$$C := (X^2 + 3*X + 2)/(X^2 + 4*X + 3)$$

The common factor X+1 wasn't cancelled. NUM doesn't do any better:

```
NUM(C);
```

$$X^2 + 3*X + 2$$

The ON GCD command, one of the many commands to be discussed in Chapter 3, instructs REDUCE to work harder and cancel all common factors:

```
ON GCD$

C;
```

$$(X + 2)/(X + 3)$$

To return REDUCE to normal mode:

```
OFF GCD$

C;
```

$$(X^2 + 3*X + 2)/(X^2 + 4*X + 3)$$

Note that after the OFF GCD the form of C is the same as it originally was, since after all we didn't change its assigned value. If we wanted to preserve the simplified form it had while GCD was ON, we should have used `C:=C` and not just C:

```
ON GCD$

C:=C;
            C := (X + 2)/(X + 3)
OFF GCD$

C;
            (X + 2)/(X + 3)
```

Exercise 1.7.1.

If N is an integer, what are the possible values of DEN(N/2)? What do they indicate?

Exercise 1.7.2.

Let Z be the numerator of the difference between

$$\frac{X+3}{(X+4)(X-7)(X-8)} \quad \text{and} \quad \frac{A}{X+4} + \frac{B}{X-7} + \frac{C}{X-8}.$$

Print out the results of substituting in Z the three values -4, 7, and 8, one after another, for X. Give A, B, C the values they need to have in order to make each of the three results zero. Then check that Z is now zero. (This is one approach to the technique of Partial Fractions usually taught in Calculus courses as a step in the integration of rational functions. <u>Caution</u>: More complicated examples, in which the denominator involves factors like (X + SQRT 5) and (X - SQRT 5), may require this calculation to be done in the ON GCD mode which was just discussed.)

1.8. Differentiation

The function DF performs differentiation of expressions.

Differentiation

```
EE:=X**3;
```

$$EE := X^3$$

```
W:=DF(EE,X);
```

$$W := 3*X^2$$

The symbol DF(EE,X) is read "derivative of EE with respect to X". The first argument of the DF(EE,X), the EE, was evaluated and was found to be X**3. The second argument, X, should be clear. The DF operation calculates what is normally written as $d(X^3)/dX$ in Calculus.

DF knows most of the rules of differentiation of continuous functions, including many trigonometric forms.

Exercise 1.8.1.

Use REDUCE to find dY/dX where Y is X/((X+1)*(X+2)). Then examine the answer (by eye, not by REDUCE) to find where the derivative is zero.

(This exercise is in part intended to illustrate that REDUCE doesn't have to be asked to do *every* aspect of a calculation in order to be worth while!)

Exercise 1.8.2.

If the second argument in DF(EE,X) is not clear, it also is evaluated and the differentiation is with respect to the result. Demonstrate this. Also note the error message obtained if the value of the second argument is not just a variable but a more complicated expression.

DF can be used to find higher derivatives in a single step, by introducing an integer-valued third argument. To find the second derivative of EE:

```
S:=DF(EE,X,2);

     S := 6*X
```

The DF operation is actually partial differentiation. Variables other than the second argument are (normally) treated as if they were constants.

```
DF(X*Y,X);

     Y
```

If one variable, Y, is to be treated as depending on another, X, the declaration

```
DEPEND Y,X$
```

must first be entered. With it, the result is quite different:

```
A:=DF(X*Y,X);

     A := DF(Y,X)*X + Y
```

Notice that the answer contains the symbol DF(Y,X). For the time being this symbol is carried along "as is", as an unevaluated kernel. If Y is eventually assigned a value (instead of being clear as is now the case) the symbol will be evaluated:

```
Y:=X**10$

A;
            10
        11*X
```

If later in the REDUCE session Y is no longer to be treated as if it depended on X when clear, the command NODEPEND Y,X$ should be used to nullify the DEPEND declaration.

Exercise 1.8.3.

Demonstrate that REDUCE knows the product rule of differentiation: that the derivative of U*V with respect to X is DF(U,X)*V + DF(V,X)*U. Don't just try an example; show that it knows it as a general rule. Hint: use DEPEND.

1.8.1. Partial derivatives

An expression can be differentiated with respect to more than one variable. The most general form of DF is

    DF(EE,X,2,Y,3,Z,5,...)

meaning second derivative with respect to X, third with respect to Y, etc. If a number is omitted after a variable, a 1 (first derivative) is assumed. For example

    DF(Y,X,2,U,W,2)

means $d^5 Y/dX^2 dU dW^2$.

If Z depends on both X and Y, so DF(Z,X) and DF(Z,Y) are both to be left as is, instead of giving two separate DEPEND commands we can write:

    DEPEND Z,X,Y$

Note that DEPEND Z,X,Y means that Z depends on X and Y, not that Z and X both depend on Y. To express the latter, two DEPEND commands are needed.

1.9. Integration

A symbolic integration operation INT is available. In some implementations of REDUCE, the command **LOAD "INT"$** or **LOAD "INTEG"$** must be given before the first use of INT in a session.

    INT(EE,X)

has as result the indefinite integral of the expression EE with respect to the variable X. The arbitrary constant is not represented.

```
INT(X**10,X);

        11
       X  /11
```

As is well known, not every integration problem has an answer which can be expressed in terms of combinations of familiar functions. For example, the calculations `INT(E**(X**2),X)` and `INT(SIN(X)/X,X)` can not be expected to succeed. But we must warn that REDUCE's INT operator will not always find the answer even if there is an answer which could be represented in terms of standard functions. One of INT's limitations is that it doesn't handle integrals like `INT(1/SQRT(1-X**2),X)` which can in fact be evaluated in terms of the ASIN (arcsine) function. This inability results from some gaps in the theory on which the integration method used by REDUCE is based, but may possibly be repaired in the future.

In some implementations of REDUCE, one must first make sure there is enough organized memory space ("core") for INT to function. For example, on the DEC 20 it is advisable to issue the command

CORE 70$

before trying to use INT for the first time in a run. Other implementations, on the other hand, may not even have a CORE command.

We give an example showing the ability of the integration operation to do factoring and partial fractions decomposition when necessary:

```
A:=(X+1)*(X+5)**2*(X+7);

            4       3        2
     A := X  + 18*X  + 112*X  + 270*X + 175
```

(The value of A is stored in this multiplied-out form. INT will have to recover the factors by factoring!)

```
INT(X/A,X);

    (28*LOG(X + 7)*X + 140*LOG(X + 7) - 27*LOG(X + 5)*X

    - 135*LOG(X + 5) - LOG(X + 1)*X- 5*LOG(X + 1)

    + 12*X)/(96*(X + 5))
```

When INT fails to carry out an integration it is asked to do, it responds in one of two ways:

1) It returns the input, INT(... , ...) unchanged.

2) It returns an expression involving INT of some other expression (unfortunately sometimes more complicated than the original one).

The DF operator knows that DF of an INT is the expression with which you started. For example, consider

```
A := INT(SIN (10*X)/X, X);

    A := INT(SIN(10*X)/X,X)
```

This is an instance of case 1: the answer is the problem unchanged, since $SIN(10*X)/X$ can not be integrated. But we can differentiate A to get $SIN(10*X)/X$ back:

```
B := DF(A,X);

    B := SIN(10*X)/X
```

The reader who is curious about the process INT goes through in attacking a problem can eavesdrop on it by typing the command ON TRINT$ before asking INT to find some integrals. Once you've seen the trace of enough integration, use OFF TRINT$ to disable this extremely verbose mode.

1.10. FOR ... DO

It is possible to ask REDUCE to perform some operation or group of operations repeatedly. We begin with a simple example:

```
A := 0$

FOR I:=1:100 DO A:=A+I**2$
```

This will cause A:=A+I**2 to be performed repeatedly, with I taking on the values 1,2,3, ... 100 in turn. Since A started with the value 0, this will set A to the sum of the squares of the integers from 1 to 100.

(What happened to the restriction that I can not be changed from its standard meaning of SQRT(-1)? The answer is that this restriction is suspended if I is used as the "controlled variable" in a FOR statement. E can be used similarly, in spite of E being otherwise a reserved variable. But T can not be so used. We will see later that this same lifting of the restriction applies if I or E is used as a formal parameter in a procedure.)

The FOR ... DO doesn't have a result which prints automatically, even if a semicolon follows it. If we want to see what the final value of A is, we need to follow this with

```
A;
        338350
```

If we wanted to add the squares of 1,3,5, ... 99 instead, we could use

```
A:=0$

FOR I:=1 STEP 2 UNTIL 99 DO A:=A+I**2$

A;
        166650
```

In the first example, with FOR I:=1:100, the colon between the 1 and the 100 is an abbreviation REDUCE recognizes to mean

FOR ... DO   33

STEP 1 UNTIL.

Warning to Pascal programmers: If you attempt to use the word "TO" instead of a colon, as for example in Pascal's FOR I := 1 TO 100 DO ..., you'll get the misleading error message

```
***** DO invalid in FOR statement
```

For the record, we mention that STEP S UNTIL L is legal even with variables for S and L (and changed in the DO part), and even with S negative. Whenever S is positive, the FOR checks that the controlled variable is less than or equal to the limit value; whenever S is negative, that the variable is greater than or equal to the limit value. But while changing S and L inside the loop is permitted, it is not a recommended programming technique!

If we wanted to add, and call A, the sums of the squares of 1,2,3, ... 100, and at the same time wanted B to be the sums of their cubes, we could write

```
A:=0$ B:=0$

FOR I:=1:100 DO <<A:=A+I**2$ B:=B+I**3>>$
```

When more than a single statement is to be controlled by the FOR ... DO, the statements must be enclosed between the statement grouping symbols << ... >>. Many people use the << ... >> even when a single statement is being controlled, to make the program easier to read:

```
FOR I:=1:100 DO <<A:=A+I**2>>$
```

We happened to use the letter I as the controlled variable in all our examples of FOR ... DO. Of course any other variable could be used equally well, here and in the FOR ... SUM and FOR ... PRODUCT discussed below. As we said before, the variable so used is a "local" variable. It should be thought of as I OF THE FOR STATEMENT rather than as I. This is why I doesn't act as a reserved variable. For the same reason, any value the controlled variable had before the FOR is carried out remains unchanged:

```
K:=X+Y+Z;
```

$$K := X + Y + Z$$

```
A:=0$
FOR K:=1:100 DO A:=A+K**2$
K;
```

$$X + Y + Z$$

because it was K OF THE FOR STATEMENT that was changed.

Exercise 1.10.1.

Try the following, and explain the answer you obtain:

```
EE := N**2$
A := 0$
FOR N := 1:10 DO A := A+EE$
A;
```

1.11. FOR ... SUM

There is an easier way to ask REDUCE to set A equal to the sum of the squares of 1,2,3, ... 100:

```
A:=FOR I:=1:100 SUM I**2;
```

The FOR ... SUM corresponds to the large capital Sigma sign of algebra. FOR ... DO was a Statement (it did something, but had no resulting value of itself); FOR ... SUM is an Expression (it has a value which can be used in an assignment).

Another example:

```
B:=FOR I:=1:7 SUM X**I;
```

$$B := X + X^2 + X^3 + X^4 + X^5 + X^6 + X^7$$

This shows that what FOR ... SUM adds need not be numbers, but can be arbitrary expressions. Still another example:

```
C:=100*(FOR I:=1 STEP 2 UNTIL 99 SUM I**2)**3;
```

C is set equal to 100 times the cube of the sum of the squares of the numbers 1,3,5, ... 99. This shows a FOR ... SUM used as a component of a larger expression. Note the parentheses around it. Without the parentheses, the values of I**2**3 would have been summed, since SUM sums the longest possible expression which follows. As a simple example,

```
FOR I:=1:10 SUM A + B;
```

evaluates to 10*(A + B), not to 10*A + B.

Since FOR ... SUM is an expression and has only a single value, we obviously can not use it to form two different sums simultaneously the way we could with FOR ... DO.

Exercise 1.11.1.

Calculate $(1-X)^2 * (1 + 2*X + 3*X^2 + ... + 10*X^9)$ using a single REDUCE command.

Exercise 1.11.2.

Convert Exercise 1.10.1 to the FOR ... SUM form, and try it. Do you get the same unexpected answer?

Since FOR ... SUM ... is an expression, it can itself be used in a FOR ... SUM ... expression. Of course different controlled variables have to be used for the two FOR's:

```
DS := FOR I:=1:10 SUM
         FOR J:=1:5 SUM A(I,J) $
```

would set DS to $A(1,1) + ... + A(10,5)$. The two FOR's are written on separate lines only for ease of reading.

(We don't expect the reader to understand, at this point, what the notation $A(I,J)$ means. He can think of it as a shorthand for some expression involving I and J. Later he will learn that A here could be a procedure name, an array name, a matrix name, or an operator name!)

1.12. FOR ... PRODUCT

The reader who understands FOR ... SUM will require no long explanation of FOR ... PRODUCT. FOR ... PRODUCT corresponds to the large capital Pi symbol of algebra, and computes the product of values formed one by one.

```
C:=FOR I:=1:100 PRODUCT I;
```

multiplies together all the integers 1,2,3, ... 100 and sets C equal to their product, 100 factorial.

```
D:=FOR I:=0:4 PRODUCT (X+I*Y);
```

is a quick way to write

```
D:=X * (X+Y) * (X+2*Y) * (X+3*Y) * (X+4*Y);
```

Exercise 1.12.1.

Verify by a REDUCE computation, followed by inspection, that the product $(X+5)(X+4)(X+3)...(X-4)(X-5)$ has no even powers of X.

1.13. WHILE ... DO

The FOR ... DO feature allows easy coding of a repeated operation in which the number of repetitions is known in advance. If the criterion for repetition is more complicated, WHILE ... DO can often be used. We start with a numerical example.

Suppose we want to add up a series of numerical terms,

generated one by one, until we reach a term which is less than 1/100 in value. For our simple example, let us suppose the first term equals 1 and each term is gotten from the one before by taking one third of it and adding one third its square. We would write:

```
EE:=0$ TERM:=1$

WHILE TERM >= 1/100 DO
        <<EE:=EE+TERM; TERM:=(TERM + TERM**2)/3>>$

EE;
            1417919029503445/617673396283947
```

If we prefer a decimal approximation:

```
ON FLOAT;

EE;
       2.2955805

OFF FLOAT;            % Back to normal mode!
```

As long as TERM is greater than or equal (>=) to 1/100 it will be added to EE and the next TERM calculated. As soon as TERM becomes less than 1/100 the WHILE test fails and the loop exits.

The WHILE ... DO controls the single statement following DO. If several statements are to be repeated, they must be grouped using the << ... >> as in the example.

The WHILE condition is tested each time <u>before</u> the action following the DO is attempted. If the condition is false to begin with, the action is not performed at all. Make sure that what is to be tested has an appropriate value before starting the loop.

1.14. Boolean expressions

Built into REDUCE are certain functions and operations which deliver as result the truth values true or false. They are intended for use, often in combination, for specifying the test part of a WHILE statement and in the forthcoming REPEAT, IF, and FOR ALL ... SUCH THAT ... LET statements. Among these operations are the following comparison operations which work only on numbers, or expressions whose values are numbers (i.e., contain no variables or the like):

```
>       greater than
>=      greater than or equal
<       less than
<=      less than or equal.
```

We point out that in earlier versions of REDUCE these operations worked only on integers, not on fractions. In these versions, if we wanted to compare two numbers X and Y, and didn't know whether they were integers, we had to instead compare the numerator of their difference, NUM(X-Y), with 0. (The numerator is, of course, always an integer.)

The operators

```
=       equal
NEQ     not equal
```

work not only for numbers, but for arbitrary algebraic expressions. (Pascal programmers note: "<>" can not be used for NEQ.)

For combining truth values we can use

```
        AND
        OR
and     NOT.
```

The notation $5 < X < 1000$ is not allowed. One must use $5 < X$ AND $X < 1000$.

Parentheses can be used to indicate grouping:

(X>5 AND X<1000) OR (X>2000 AND X<3000)

is true if the integer X is either between 5 and 1000 or between 2000 and 3000.

Of these three operations, NOT has the highest priority, followed by AND and then by OR. It follows that the parentheses above are not actually needed. But it is advisable to include them anyway, for clarity.

A Boolean function of occasional use is NUMBERP.

NUMBERP X

is true if X is a number, and false otherwise. (In some older versions of REDUCE, NUMBERP was false for a number like 2/3, which was considered to be the quotient of two integers, not a single number.)

SQRT 2 and similar expressions are not considered to be numbers by NUMBERP, unless the mode ON BIGFLOAT$ ON NUMVAL$ (see page 159) has been set in order to replace the "SQRT 2" form by a numerical approximation.

The principal use of NUMBERP is to prevent trouble in situations where a comparison like X<Y is to be made when it is not certain X and Y are actually numbers (and not, for example, clear).

NUMBERP X AND NUMBERP Y AND X<Y

is true if X and Y are both numbers and X is less than Y. If either X or Y is not a number, this Boolean expression evaluates to false instead of terminating with an error message.

Note that the order of the parts connected by AND is critical. The parts are tested one by one, in the order they appear, until one is found which is not true, or all have been tested and found true. If, say, NUMBERP X is true then NUMBER Y is tested. If it is false then the remaining part, X<Y, is not evaluated, so no error condition arises. If the X<Y test were first the NUMBERP tests would be too late and serve no purpose.

(The analogous remark applies to parts of a Boolean expression separated by OR. As soon as a true part is encountered, while testing the parts one-by-one from left to right, the remaining parts are not evaluated.)

We had said that the result of evaluating an expression is

automatically printed out if a semicolon is used as terminator. This is not true for Boolean expressions. The value, true or false, of a Boolean expression can not be printed directly by simply entering a "statement" like

    A=B;   or   A>B;   or   NUMBERP A;

If you try it, you just get an error message like "= invalid as algebraic operator". Anticipating the IF ... THEN statement to be formally introduced later (page 113), we note that it can be used to print out the value of a Boolean expression:

    X := 123/7890 $

    Y := 25/152 $

    IF X>Y THEN TRUE ELSE FALSE;

        FALSE

1.15. REPEAT ... UNTIL

REPEAT ... UNTIL is very similar in purpose to WHILE ... DO. There are two essential differences:

1) In REPEAT ... UNTIL the test is performed <u>after</u> the controlled statement (or group << ... >> of statements) is executed, so the controlled statement is always executed at least once.

2) In REPEAT ... UNTIL the test is a test for when to stop rather than when to continue, so its "polarity" is the opposite of that in WHILE ... DO.

We rewrite the example from the WHILE ... DO section:

```
EE:=0$ TERM:=1$

REPEAT <<EE:=EE+TERM; TERM:=(TERM + TERM**2)/3>>
       UNTIL TERM < 1/100;

EE;
        1417919029503445/617673396283947
```

Note to Pascal programmers: the syntax of REDUCE's REPEAT statement differs from Pascal's in that only a single statement is allowed between the REPEAT and the UNTIL in REDUCE. If several statements are required, they must be grouped, using << ... >> in the usual way. (In Pascal, REPEAT and UNTIL are themselves treated as grouping symbols.)

2. A Harder Look

This chapter introduces numerous features of REDUCE which are not as elementary as those presented in the previous chapter. Not every REDUCE user will require these additional features, which is fortunate because some of them -- such as the concept of the "substitution environment" -- involve logical intricacies.

2.1. The SUBSTITUTION function

The SUB function gives the result of replacing every occurrence of a variable in an expression by something else -- another variable or an expression.

    EE:=X**2 + Y**2;

$$EE := X^2 + Y^2$$

    F:=SUB(X=A+B,EE);

$$F := A^2 + 2*A*B + B^2 + Y^2$$

F has been obtained from EE by substituting A+B for every X in the value of EE.

SUB can be used to make any number of substitutions simultaneously. The "variable=replaced by" pairs are written first, and the expression in which the replacements are to be done is last. All are separated by commas.

    G:=SUB(X=A+B,Y=Y+1,EE);

$$G := A^2 + 2*A*B + B^2 + Y^2 + 2*Y + 1$$

This time X was replaced by A+B and Y by Y+1. EE itself remains unchanged, and X, Y remain clear:

The SUBSTITUTION function

```
EE;
```

$$X^2 + Y^2$$

These examples assumed that the variables for which we are making substitutions are clear. That is the normal use of SUB, because after all if the variable X isn't clear, then there can't be any X in the current value of the expression to be modified. Just for the record, if X is not clear but has as value some other variable, say P, then SUB(X=...,EE) is executed as if it said SUB(P=...,EE). If X has as value an expression like P+Q, not a variable, then SUB(X=...,EE) produces the error message "X invalid as kernel".

The reader may have been bothered about the substitution Y=Y+1. This doesn't lead to an endless evaluation loop, the way an assignment Y:=Y+1 would (assuming Y clear), because SUB makes only a single sweep through the expression. In greater detail:

SUB first simplifies the expression; then replaces in it the variables occurring on the substitution list; and finally resimplifies the result. The substitutions are made only once, so no endless loops occur. No assignment like X:=A+B actually takes place, so X remains clear.

One more example:

```
EE:=2*X + 3*Y;

    EE := 2*X + 3*Y

EE:=SUB(X=Y,Y=X,EE);

    EE := 3*X + 2*Y
```

Here SUB was used to interchange X and Y in an expression.

Exercise 2.1.1.

Find a way to interchange X and Y in an expression without using SUB, only using assignments and CLEAR. (One or more intermediate variables, that is variables that do not occur in the expression, will probably be needed.)

Exercise 2.1.2.

(For readers who remember Calculus) Find the Maclaurin expansion of $1/((X-1)*(X-2))$ in powers of X up to and including the $X**5$ term. (You may know this as the Taylor expansion at 0.) Hint: The formula for the Maclaurin expansion, as found in Calculus books, includes expressions like $f'''(0)$, which could be represented in REDUCE as SUB(X=0,DF(F,X,3)). You will want to use this, or something similar, in a FOR statement of some kind. The factorial which is part of the formula could be represented as a FOR ... PRODUCT. Challenge: Now solve the problem in a way which will execute more efficiently, by not finding each higher derivative and each factorial from scratch. (Use variables to represent the several factors of a term, and figure out how to get the factors for each term from those of the preceding.)

Exercise 2.1.3.

Write a single expression, using SUB, whose value is 0 if the expression A doesn't contain the variable X, and different from 0 if it does. Does your method work no matter what A is? (This problem is discussed in the Case Studies chapter, starting on page 210.)

Exercise 2.1.4.

(This is mathematically interesting but quite simple.) Suppose the expression A is known to be the product of some expression F that contains only the variable X, and some expression G that contains only the variable Y. Show how to use SUB to find F and G given only A. (Note that the answer isn't unique. If A is $X^2Y^3/10$ then we could decompose it into $F=X^2$ and $G=Y^3/10$, or $F=X^2/10$ and $G=Y^3$, or even $F=7*X^2/15$ and $G=3*Y^3/14$.) Hint: If you pick any numerical value for Y and substitute it in A, you get a possible expression F, unless the value you picked happens to make A equal to zero.

2.2. ARRAY and OPERATOR

The words "array" and "operator" have very different connotations for most programmers, but in REDUCE they stand for a pair of closely related notions. If A is either an array or an operator, a symbol like A(N) is meaningful. There are many situations in which it is not obvious whether it is more appropriate to declare a particular symbol to be an array name or an operator name.

Exercise 2.2.1.

While reading the two subsections below, list the differences between arrays and operators in tabular form. Then run test examples on the computer to check that your table is correct. (After you study matrices in REDUCE you will be asked to add a third column to this table of comparisons.)

2.2.1. Arrays

If ten related expressions (or variables or numbers) have to be handled by REDUCE, it may be unnatural or unreasonable to invent ten different names for them. It is surely preferable to be able to deal with them in a uniform way. The Array concept allows us to refer to them systematically as, say, A(0), A(1), A(2), ..., A(9).

Just because the symbol A(0) is available for use does not mean that it <u>must</u> be used. In some situations it's more natural to start counting from 0; in other situations, from 1. Let us assume now that we don't want to use A(0), and instead let the numbering run A(1) through A(10).

If the symbol A is to be used this way, it is necessary to first declare this fact, and to indicate what the largest place number -- here 10 -- might possibly be. Array declarations in REDUCE are similar in appearance to the DIM or ARRAY declarations of many other languages:

```
ARRAY A(10),B(25);
```

or, on separate lines,

```
ARRAY A(10);
ARRAY B(25);
```

(A, B must not currently have values assigned to them, as ordinary variables, if the declarations are to be accepted. If an assignment like `A := P;` has been entered, and A has not been cleared since, then attempting to declare A to be an array leads to a "VARIABLE A invalid as ARRAY" printout.) From this point on, the symbols A(0), A(1), ..., A(10), and B(0), B(1), ..., B(25), can be used the same way that simple variables like P, Q, or ELEPHANT are used:

```
A(10) := (P+Q)**2;

A(7) := 7*ELEPHANT;

X := A(10) - A(7);

B(10) := P-Q;
```

We call A(0), ..., A(10) the <u>places</u> or <u>positions</u> of the array A, and the number 10 is called its array bound. However by tradition it is generally referred to as the dimension of the array. Note that, since array index 0 is allowed, the array can hold one more item than its "dimension" would indicate.

Instead of supplying an explicit number we can write an expression when defining the array bound, provided the value of the expression is a positive integer:

```
ARRAY C(5*X+Y);    % means ARRAY C(57) if X=10 and Y=7
```

Similarly, in referring to an array position we can use a symbol or expression instead of an explicit number, provided the symbol or expression has a numerical value and that value is a positive (or zero) integer no greater than the array bound that was declared. For example, if X=10, Y=7, and Z is clear, then we can speak of A(X), B(Y), and A(X-Y),

        but not of A(X/Y)   [X/Y = 10/7, not an integer];

        not A(Z)   [Z = Z, not even a number];

ARRAY and OPERATOR

    not A(X+Y)   [X+Y = 17 which exceeds the array bound].

In REDUCE it is not possible to copy an array all at once. If A is an array of dimension 10, and B is of the same or larger dimension, to copy A to B a loop is necessary:

```
FOR I:=0:10 DO B(I):=A(I)$
```

An assignment B:=A would copy the value of the ordinary <u>variable</u> A, which is not related to the <u>array</u> A, to the ordinary variable, not array, B. (While only clear variables can be made into arrays, once a variable is declared to be an array it <u>can</u> also be assigned a value as an ordinary variable. But it is unwise to do this, to use one symbol with two meanings.)

Exercise 2.2.2.

Declare an array A of dimension 10, and use a FOR ... DO statement to set A(0) to 1, A(1) to X+Y, A(2) to (X+Y)**2, ..., A(10) to (X+Y)**10. Then spot check a few of these array places to see if you succeeded.

We can declare and use two- and higher-dimensional arrays:

```
ARRAY D(5,10);
ARRAY EE(1,1,1,1,1);
```

These declarations allow reference to 66 places D(0,0) through D(5,10) and to 32 places EE(0,0,0,0,0) through EE(1,1,1,1,1).

Notice that here we use the word "dimension" in a different sense! The context will normally make clear which is the intended meaning. For example, arrays such as those declared by writing ARRAY A(10) will be called one-dimensional (and that one dimension equals 10).

If D is a 2-dimensional array of dimensions (5,10), and EE is another 2-dimensional array of the same or larger dimensions, to copy D to EE a double loop is necessary:

```
    FOR I:=0:5 DO
      FOR J:=0:10 DO EE(I,J):=D(I,J)$
```

If the array is one-dimensional, and the bound or subscript a non-negative number or a variable (not a more general expression), the parentheses may be omitted both in the declaration and in the use:

```
    ARRAY A X, C 57;      % recall that X = 10

    A Y := ELEPHANT;      % recall that Y = 7

    A 9 := 5*A Y + 1;     % meaning  5 * A(7) + 1
```

As these examples show, if the parentheses are omitted there has to be a space (more exactly, at least one space) between the array name and the bound or subscript. The symbol A7 represents an ordinary variable, not related to the array place A 7.

Array declarations can be made at any time during a terminal session, as long as each occurs before its use. They do not have to be grouped together at the beginning as is required in some other languages.

If A is an array, the command CLEAR A tells REDUCE to forget that fact and delete all the values stored in all the array places. Re-declaring the array, with say ARRAY A 10, also serves to delete all the values previously stored (even if the array bound, 10, is the same as it was before); but a message "ARRAY A REDEFINED" is printed which may be annoying if the re-definition is intentional, so a CLEAR A is recommended before the array redeclaration.

We note here that once a symbol is declared to be an array name, it can not also be used to name an operator or a procedure unless its array declaration is first cancelled by using CLEAR. As was remarked before, it can still be used as a simple variable, although the wisdom of this is questionable. (It can cause trouble if you call a procedure one of whose arguments is supposed to be the name of the array, and that name also has a value as an ordinary variable.)

ARRAY and OPERATOR

All places of an array are initialized to 0 when the array is declared. In fact, array places can never be clear in the sense in which simple variables can be clear, so the result of asking for the value of an expression can never contain an array reference like A(7). If A(7) has never been assigned a value, its value is 0, not A(7).

We have already explained what a command such as CLEAR A does if A has been declared to be an array. The effect of trying to CLEAR an individual array place, by issuing a command like CLEAR A 7, is quite different, and rather unexpected but sometimes useful.

- If the current value of A(7) is a variable, say P, and P is not clear, then CLEAR A 7 actually clears P, and doesn't affect the contents of A(7) itself.

- If the current value of A(7) is a variable, say P, and P is clear, then CLEAR A 7 results in an error message, "P not found".

- If the current value of A(7) is a number, say 123, CLEAR A 7 results in an error message, "Substitution for 123 not allowed".

- If the current value of A(7) was zero, the error message was, in older versions of REDUCE, sometimes given as "NIL is a reserved identifier".

Exercise 2.2.3.

Test the above assertions, describing the effect of CLEAR A 7, in the version of REDUCE you are using. (In some older versions of REDUCE, CLEAR A 7 simply set A(7) to zero.)

There is one remaining case. If the value of A 7 is an expression, like U+V, the effect of CLEAR A 7 is the same as if CLEAR U+V had been entered. The meaning of this will be explained in the section on Let and Clear that begins on page 77.

2.2.2. Operators

The declaration

OPERATOR H,G1,ARCTAN;

allows symbols like H(W), H(X,Y,Z), G1(P+Q), ARCTAN(U/V) to be used in expressions. No meaning or properties of the operator are implied. There are no restrictions on the type of value, if any, the expression or expressions in the parentheses have. By analogy with the terminology for arrays, we will refer to any particular symbol such as H(W) or H(X,P+Q) as an operator "place", keeping the quotes around "place" as a reminder that this usage is metaphorical.

For example, we could make an operator declaration, and two assignments

OPERATOR BABY$

BABY(CAT):=KITTEN$

BABY(DOG):=PUPPY$

and later retrieve and use the values:

5*BABY(CAT) + 2*BABY(DOG);

5*KITTEN + 2*PUPPY

As a more serious example, we could let Y be the "general form" of a quadratic polynomial:

OPERATOR A$

Y := A(2)*X**2 + A(1)*X + A(0)$

and substitute various values for X:

ARRAY and OPERATOR

X:=M$

Y;

$$A(2)*M^2 + A(1)*M + A(0)$$

X:=12$

Y;

$$144*A(2) + 12*A(1) + A(0)$$

X:=P+Q$

Y;

$$A(2)*P^2 + 2*A(2)*P*Q + A(2)*Q^2$$
$$+ A(1)*P + A(1)*Q + A(0)$$

An operator "place" is initially clear in the same sense that a simple variable is clear: its value is its own name. This is why

    Y := A(2)*X**2 + A(1)*X + A(0)$

could be used as a general formula with A an operator but would not have worked if A were an array name.

Exercise 2.2.4.

Explain the above remark.

To see the advantage of using $A(2)$, $A(1)$, $A(0)$ instead of, say, A, B, C, note that we could set Y to the general N'th degree polynomial, for any given N, by writing

    Y := (FOR I:=0:N SUM A(I) * X**I)$

If we try to use a symbol like H(W) but had forgotten to declare H as OPERATOR, the system would obligingly ask

>   **DECLARE H OPERATOR?**

to which we would reply by typing the letter Y -- meaning YES -- and striking RETURN. This would in effect supply the missing declaration. (If we had meant H to be an array instead, our only recourse would be to answer N, meaning NO, make the ARRAY declaration for H with its correct bound, and then either retype the expression containing the H(W), or use the RETRY command described in the chapter on Running REDUCE.)

A few more illustrations:

```
H(P+Q):=THIS-IS-IT$

H(-57):=HEINZ$         % we have now stored values for
                       % two of the "places" of H

Y:=P-Q$

H(Y);
        H(P - Q)       % no value was stored for H(P-Q)

H(Y+2*Q);
        THIS - IS - IT    % since Y+2*Q = P+Q

X:=77$

H(X);
        H(77)          % The X is evaluated to 77 first,
                       % so stored H(X) is not retrieved
```

The command CLEAR H(P+Q) restores H(P+Q) to its original condition, after an assignment has changed it:

```
H(P+Q);
        THIS - IS - IT

CLEAR H(P+Q);

H(P+Q);
        H(P + Q)
```

Any operator symbol can be used freely as a 1-, 2-, 3-, etc.-place operator, and no confusion arises between the different uses. Thus if H is an operator then H(P), H(P,P), H(P,Q), H(Q,P), H(P,Q,R), H(P+Q,R) are all distinct and can hold different values.

ARRAY and OPERATOR   53

H(P+Q) and H(Q+P), on the other hand, are the same, since only the value of what's in the parentheses matters, not the form in which it is given.

We can even use an operator as a zero-place operator, writing H() with empty parentheses. H() can be used to hold a value quite distinct from the value of the variable H. We don't recommend this confusing use of an operator symbol.

We note that REDUCE discourages but does not forbid a symbol being used both as an ordinary variable and as an operator symbol. If H has been assigned a value as an ordinary variable, an attempt to declare OPERATOR H results in an error message. However, once a symbol has been declared an operator, REDUCE doesn't prevent one from assigning a value to it as a variable. (The analogous fact for arrays has already been noted.)

If the operator is being used as a one-place operator, and that place is to be filled with a (non-negative) number or a variable (not a more general expression), the parentheses may be omitted. This convention is the same as for arrays. The reader will find the same convention will apply to procedures, too.

If H is an operator symbol and EE is an expression in which it is used, the statement

```
F := SUB(H=Q,EE)$
```

attempts to set F to the result of changing every H(...) in EE to Q(...). If Q is an operator already, or can be declared to be an operator, there is no trouble. If Q is the name of an array, the resulting parts of the form Q(...) in EE are treated as array place references: if the value of the (...) is an integer within the array bounds of Q, the value stored in that array place is substituted for the Q(...); if not, "subscript out of range" or "invalid as integer" error messages would result. If Q is a number or an expression that is not a single variable, an "invalid as operator" message results.

If the speed of execution is an issue, it should be noted that retrieving the value stored in an operator "place" requires a sequential search through all the values stored, starting with the most recent. Retrieving the value stored in an array place is much faster, in general.

If an operator is re-declared as operator, nothing happens except that a warning message is issued. If H is an operator name, CLEAR H causes REDUCE to forget that fact, and then H could be reused in other ways, for example as an array name. If H(ABC) had been assigned the value 123, that fact is apparently forgotten. However, if later H is re-declared to be an operator, all the previously stored "place" values are suddenly remembered again! (Versions of REDUCE before REDUCE 3.2 behaved differently in this respect.)

Since CLEAR H, if H is an operator, does not delete the values stored for operator "places", if they are to be deleted the programmer must remember what "places" have been set, and clear them individually, by a series of commands such as **CLEAR H(ABC)**$.

To give a meaning to an operator symbol, or to express some of its properties, the LET statements explained in a later section can be used. It may develop that the symbol shouldn't have been declared to be an OPERATOR at all, but defined to be a PROCEDURE. (That's a quite different thing!)

If an operator symbol is later used as a procedure name, the values that may have been assigned to an operator "place" are superseded by the values computed by the procedure.

2.3. Matrices

A matrix in REDUCE is superficially like a 2-dimensional array. The symbol A is declared to represent a matrix of 3 rows and 4 columns by a declaration

MATRIX A(3,4)$

This allows one to use the 12 symbols A(1,1) through A(3,4) the same way one would use array positions. (Note that subscripts of 0

Matrices

are not permitted.) Just as for arrays, these places are all initially zero, and the command CLEAR A causes the matrix declaration (and the matrix entries) to be forgotten.

Several matrices can be declared at once:

`MATRIX A(3,4), B(5,5), C(5,5)$`

If we want the square matrix B to contain zeros except for the number 1 everywhere on the diagonal (i.e. if we want it to be the 5 by 5 identity matrix) we need merely follow the declaration by

`FOR I:=1:5 DO B(I,I):=1$`

relying on the initialization of the matrix entries to zero.

Exercise 2.3.1.

Use at most two FOR ... DO statements to define B to be the matrix

$$\begin{Vmatrix} 2X & 1 & 0 & 0 & 0 \\ 0 & 2X^2 & 1 & 0 & 0 \\ 0 & 0 & 2X^3 & 1 & 0 \\ 0 & 0 & 0 & 2X^4 & 1 \\ 0 & 0 & 0 & 0 & 2X^5 \end{Vmatrix}$$

(Do not assume, from seeing the way this problem is stated, that REDUCE prints out matrices in such a rectangular pattern. Printing matrices will be discussed a few pages later; meanwhile, consider that the expressions that are the entries of realistic matrices may not be as simple as $2X^n$ but may each take many lines to print!)

REDUCE doesn't have any special provision for matrices that have a single column or a single row. They are treated simply as

N-by-1 or 1-by-N matrices. The word VECTOR is a reserved word in some versions of REDUCE but has an entirely different meaning there.

So far, matrices and two-dimensional arrays seem to be very similar, except that matrices don't have a row or column numbered zero. The first difference we note, a rather minor one, is that the := operator can copy matrices "all at once":

```
C := B$
```

copies the 25 elements of the matrix B into the matrix C. If B and C were arrays this would not work, since array names are simultaneously names of simple variables, and C := B would just copy the value of the simple variable B to the simple variable C, and not affect the so-named arrays at all. (Don't attempt to use LET statements, which will be introduced later in this book, to copy matrices. LET C = B appears to work, but actually links B and C in such a way that if an entry in B or in C is changed subsequently, the corresponding entry in the other matrix changes also!)

Another minor difference is that if A is a matrix then the command CLEAR A(2,3) resets A(2,3) to zero. Recall that if A were an array then CLEAR A(2,3) would attempt to "clear" whatever was the prior value of A(2,3); and if A were an operator then CLEAR A(2,3) would reset A(2,3) to the symbolic form A(2,3).

The declared matrix size plays a minor role, in contrast to the importance, for arrays, of the declared bounds. Even with the declarations given above, the assignment

```
A := B$
```

is entirely legal! Since B is 5 by 5, the matrix bounds of A are changed to 5 by 5 and the matrix B is then copied to A.

In fact the matrix size need not even be given in the declaration. If it's given, the matrix is at least temporarily of that definite size. But the declarations

Matrices

    MATRIX D,EE,F$

are perfectly valid, declaring D, EE, and F to be matrix variables but of indefinite size. The matrix size of any of these, say D, only becomes set when a matrix of definite size is copied to D using the := assignment operation. The size then remains set until changed by another matrix assignment.

There are only two disadvantages of the simpler declaration: D is not initialized to zero; and one can't set individual places in D the way we did in B, until the matrix size is set.

The matrix declaration for D is not even absolutely necessary. If an assignment

    D := A$

(with A as a matrix) is attempted when D has not yet been declared to be a matrix, one needs only to reply Y to the question

    DECLARE D MATRIX ? (Y/N)

for the declaration to be supplied by REDUCE.

Sometimes we want to develop a general formula in which the entries of a matrix are represented by symbols such as $A(1,1)$, $A(1,2)$, and so on. In this case we should not declare A to be a matrix. A should be declared to be an operator, and another but related symbol, such as AA, should be declared to be the matrix, and filled with the entries $A(1,1)$, etc. For example, if we want a 2 by 2 general matrix, we may write

    OPERATOR A$

    MATRIX AA(2,2)$

    FOR I:=1 : 2 DO
       FOR J:=1 : 2 DO
          AA(I,J) := A(I,J)$

Now we can, for example, print the general formula for the determinant of 2 by 2 matrices, using the matrix operation DET that will be introduced formally a few pages hence:

    DET AA;

        A(2,2)*A(1,1) - A(2,1)*A(1,2)

2.3.1. MAT

Matrices can be defined element by element by individual assignments like

    A(1,1) := ... $

or all at once by using the matrix constructor operator MAT. A matrix like

$$\begin{Vmatrix} 1 & 2 & 3 \\ 4 & 5 & 6 \\ P & Q & R \\ U & V & W \end{Vmatrix}$$

can be written in the form

    MAT((1,2,3),(4,5,6),(P,Q,R),(U,V,W))

Since REDUCE doesn't care about the division of an expression into lines, but the human eye does, this could be better typed as

    MAT(
    (1,2,3),
    (4,5,6),
    (P,Q,R),
    (U,V,W))

Note the commas between the elements of each row, the parentheses around each row, and the comma after each row except the last. The last row is followed instead by a second parenthesis.

If we wanted to set the matrix A to this value, we need merely write an assignment statement

```
A := MAT(
        (1,2,3),
        (4,5,6),
        (P,Q,R),
        (U,V,W) )$
```

Of course if P,Q, and so on are not clear, it is their values that will be entered into the matrix. Expressions could also be written directly between the parentheses and commas of the MAT expression.

We don't recommend using MAT to input large or complicated matrices by direct interactive typing. A single error can nullify a great deal of typing effort. MAT is best used through a text editor for creating a file for later input into REDUCE by means of the IN command. See the chapter on Running REDUCE for information on this topic.

2.3.2. Printing matrices

If a matrix name (or matrix expression, as described in the next subsection) followed by a semicolon is typed in, the entries of the matrix are printed out in a single column. Assume B is the 2-by-2 matrix

$$\begin{Vmatrix} AA & BB \\ CC & DD \end{Vmatrix} .$$

The output format is

B;

```
        MAT(1,1) := AA

        MAT(1,2) := BB

        MAT(2,1) := CC

        MAT(2,2) := DD
```

Note that the default name MAT is written on every line when a matrix <u>variable</u> or <u>expression</u> is printed. If the semicolon follows a matrix <u>assignment</u>, the name of the receiving matrix is printed instead:

```
C:=B;
```

```
        C(1,1)  :=  AA
        C(1,2)  :=  BB
        C(2,1)  :=  CC
        C(2,2)  :=  DD
```

If a large matrix with many zero entries is to be printed out, printing of the zero entries can be suppressed by setting the mode ON NERO$:

```
MATRIX B(3,3)$

FOR I:=1:3 DO B(I,I):=1$

ON NERO$

B;
```

```
        MAT(1,1)  :=  1
        MAT(2,2)  :=  1
        MAT(3,3)  :=  1
```

```
OFF NERO$
```

(The purpose of the final command is to restore REDUCE to its normal operating mode in which all matrix entries, including zeros, print.)

2.3.3. Matrix expressions

The symbols +, -, *, /, **, when applied to matrices, represent the normal operations of matrix algebra. The result of a legal matrix operation is again a matrix. Matrix expressions are built up from matrices by using these (and some other) operations, with parentheses as necessary to indicate grouping.

A+B and A-B are defined only if A and B are matrices of the same size.

A*B is defined only if A and B are compatible for multiplication, i.e. if the number of columns of A equals the number of rows of B. REDUCE knows that A*B and B*A are not generally equal, and will not replace one by the other.

X*A is also defined if X is an ordinary variable (not a matrix) and A is a matrix. It can also be written with the matrix first: A*X. Every element of A is multiplied by X.

A**N is defined if A is a square matrix and N is an integer (or a variable or expression whose value is an integer). For example A**3 means A*A*A. If N is negative, A**N represents a power of the inverse of A. For example A**(-3) means the cube of A**(-1), the inverse of A.

A/B is defined if B is a square matrix and A has the same number of columns as the side of the square B. It is interpreted as A*(B**(-1)). Note the order of this multiplication.

A/X is also defined if X is an ordinary variable (not a matrix) and A is a matrix. Every element of A is divided by X.

X/B is also defined if X is an ordinary variable (not a matrix) and B is a square matrix. It is interpreted as X*(B**(-1)). In particular, 1/B is the inverse of B.

As a final example, if A, B, and C are all 2-by-2 matrices, then

B**2*A - 3*B**(-2)*(A + C) + MAT((1,X),(Y,Z))/2

is a meaningful matrix expression (provided B has an inverse).

2.3.4. Other matrix operations

The matrix functions DET, TP and TRACE are available.

1.

DET(A), which can also be written as DET A, represents the determinant of the square matrix A. The determinant operation produces a scalar (i.e. non-matrix) result. Be cautious in its use: very small matrices can have very lengthy determinants unless there are many zeros or many closely related entries in the matrix.

Exercise 2.3.2.

Define BB to be the 4-by-4 matrix whose entries are B(1,1), B(1,2), ..., B(4,4), where B is an operator. Then calculate the determinant of B and count how many terms appear in the answer.

2.

TP A finds the transpose of the matrix A.

One use of TP is to simplify the input of a column matrix like

```
|| 1 ||
|| 2 ||
|| 3 ||
|| 4 ||
|| 5 ||
```

The direct way of defining this is by typing

    MAT((1),(2),(3),(4),(5))

The following is easier:

    TP MAT((1,2,3,4,5))

which defines the column matrix as the transpose of a row matrix.

Matrices

3.

TRACE A represents the trace of the square matrix A, i.e. the sum of its diagonal elements. The result is of course a scalar. Readers who have need for this matrix operation will have no difficulty using it.

2.3.5. A matrix example

We illustrate the use of matrices by finding the solution of a system of linear equations. We caution the reader that this is intended just as a simple example of a conceivable use of matrix operations. We are not claiming that matrix inversion provides an efficient way to solve systems of linear equations!

Suppose one has a system of n linear equations in n unknowns to solve.

```
1*X +  5*Y +   3*Z = 6,
2*X +  2*Y +   2*Z = 1,
7*X -  9*Y +  13*Z = 9.
```

In the familiar way we can convert this to a matrix problem: The coefficients of the n equations are stored in an n-by-n matrix A; the "constants" are made into an n-by-1 matrix C (sometimes referred to as a "column vector", but, as we said before, not so called in REDUCE); then the unknowns can be computed as the entries of the n-by-1 matrix X:=(1/A)*C since A*X = C. Here 1/A is of course the REDUCE notation for the inverse of A.

```
MATRIX A,C,X;

A:=MAT( (1, 5, 3),
        (2, 2, 2),
        (7,-9,13)    )$

C:=TP MAT((6,1,9))$
```

(As we remarked before, it is easier to type a 1-by-3 matrix, and transpose it to get the 3-by-1 matrix desired, than to type many more parentheses!)

```
X:=(1/A)*C;
    X(1,1) := ( - 13)/7      % value of X
    X(2,1) := 11/28          % value of Y
    X(3,1) := 55/28          % value of Z
```

The reader has hopefully been taught that solving a set of linear equations by inverting the matrix is inefficient. Matrix inversion is normally done by performing row reductions on an n-by-2*n matrix; solving a system can be done by the same row reduction operations performed on a matrix that is only slightly more than half the size, n-by-(n+1).

But "efficiency" has many meanings. Using matrix inversion involves about twice as much calculation than is necessary. But another kind of efficiency consists of using the tools at hand. Matrix inversion is an operation that is supplied in REDUCE; row reduction of a matrix is not. In this sense, the matrix inversion method is more efficient.

We present another system of linear equations to solve. This system differs from the first in that the coefficients in the problem are expressions, not just numbers. Also, we shall write the computation as a single long expression.

To find the solution of the system of equations

$$(P+Q)*X(1) + (U-V)*X(2) = Y1$$
$$(P-Q)*X(1) + LMNOP*X(2) = Y2$$

we can simply write

MATRIX X$

```
X := 1/MAT((P+Q,U-V), (P-Q,LMNOP))
            *    TP MAT((Y1,Y2));
```

Matrices   65

and get as output

$$X(1,1) := (LMNOP*Y1 - Y2*U + Y2*V)$$
$$/(P*LMNOP - P*U + P*V + Q*LMNOP + Q*U - Q*V)$$
$$X(2,1) := (- P*Y1 + P*Y2 + Q*Y1 + Q*Y2)$$
$$/(P*LMNOP - P*U + P*V + Q*LMNOP + Q*U - Q*V)$$

2.4. The COEFF function

It is often necessary to isolate the various terms of a polynomial during a calculation. COEFF is a built-in function for accomplishing this.

In REDUCE 3.2, COEFF works by spreading into an array the coefficients of the various powers of a variable. (The details are expected to change in future versions.) The present version of COEFF takes three arguments:

- the polynomial (or a variable or expression equal to it);
- the name of the variable;
- the name of the array to receive the coefficients. The declared size of the array doesn't matter.

The i'th array place receives the value of the coefficient (zero or not) of the i'th power of the variable in the expression. The array bound is increased or decreased to be equal to the degree of the polynomial, that is, the highest power of the variable present.

The COEFF function also delivers, as a result that can be printed or otherwise used, the degree of the polynomial. The degree of the polynomial is also saved as the value of the variable HIPOW!*. This information can be useful for setting up loops to print or otherwise process the array.

(If we only need to know the degree of a polynomial, it is simpler to use the function DEG than to use COEFF. If A is polynomial built from powers of a variable X, then the value of DEG(A,X) is the degree. With DEG we don't have to provide an array to receive the coefficients.)

Also available after using COEFF is LOWPOW!* which, as the name suggests, is the lowest power of the variable present. If the expression has a term in which the variable does not appear, LOWPOW!* is zero.

If the expression analysed is not a polynomial, an error message is given.

We give a simple example.

```
A:=(X + 5*W**2)**3$

ARRAY CO 10;

COEFF (A,X,CO);
```

        3   [the degree in X, printed since ";" was used in calling COEFF]

```
CO 0;
```
    $125*W^6$ [the term with no X]

```
CO 1;
```
    $75*W^4$ [coefficient of X**1]

```
CO 2;
```
    $15*W^2$ [coefficient of X**2]

```
CO 3;
```
    1   [coefficient of X**3]

Exercise 2.4.1.

Use COEFF twice to find the coefficient of $X^3 Y^8$ in

$(2*X + Y^2)^7 * (X^2 - Y^2 + Z)$.

Don't print out anything except the final answer.

2.4.1. Multi-dimensional arrays

It is also possible to have COEFF place the coefficients in a particular column of a multi-dimensional array. To do this, the third argument must not be just the name of the array, but a symbol like that for an array place except that the index for the relevant array column is indicated by an asterisk. For example,

```
ARRAY M(7,7,7);
COEFF (A,X,M(5,*,7))$
```

will cause the coefficients to be spread into M(5,0,7) through M(5,7,7).

(This feature of COEFF does not work correctly in all releases of version 3.2 of REDUCE. In some, the error message "TIMES invalid as integer" is produced, the "*" being misinterpreted as a TIMES sign.)

Exercise 2.4.2.

First set A(0), A(1), A(2), and A(3) to be four different cubic polynomials in X. (A may be an array or an operator, as you prefer.) Then use COEFF to store their coefficients in a 4-by-4 two-dimensional array G, with, for example, G(1,3) containing the coefficient of the cubic term of A(1). Spot-check G to see if it is correct, or print out all of G using a FOR loop containing the WRITE operation described in a later section.

2.4.2. Simple variable destinations

There is another form in which COEFF can be used. If the third argument (more exactly, the value of the third argument) is not an array name, the non-zero coefficients will be assigned to a series of simple variables whose names are made up of the name of the third argument followed by the corresponding power. A message is printed to inform the user of the names of the variables set.

```
A:=(X**2+Z)**3 + 123*X**50$

COEFF(A,X,W)$    % assuming W clear and not an array

        *** W50 W6 W4 W2 W0 ARE NON ZERO    % the message

W50;
         123

W6;
         1

W4;
         3*Z
```

and so on.

Exercise 2.4.3.

See what happens if you ask for COEFF(A,X,W) when W is not clear and

- the value of W is some different variable;

- the value of W is a more general expression;

- the value of W is the name of an array.

2.5. FACTORIZE

The FACTORIZE function built into some versions of REDUCE attempts, with varying success, to factor polynomials in one or more variables. (The function couldn't be called FACTOR because that name was preempted by the command FACTOR that is used for exerting a certain type of control over the output format. See page 136.)

In some implementations of REDUCE, the command LOAD "FACTOR"$ must be given before the first use of FACTORIZE in a session.

FACTORIZE

FACTORIZE takes two arguments:

- the polynomial (or a variable or expression equal to it);
- the name of the array to receive the factors. The declared size of the array doesn't matter.

Suppose we want to factor the polynomial A:

A;

$$75*(X^6 + 2*X^5*K + 5*X^5 + X^4*K^2 + 10*X^4*K +$$
$$5*X^3*K^2 + 7*X^3 + 14*X^3*K + 35*X^2 + 7*X*K^2 +$$
$$70*X*K + 35*K^2)$$

We must provide an array to hold the factors that are found, and call on FACTORIZE to fill them. We might as well dimension the array just to size 1, because the array bound is automatically increased or decreased to the required value.

ARRAY FA 1$

FACTORIZE(A,FA);

$$4$$

FOR I := 0:4 DO WRITE I,"   ",FA(I)$

```
    0      75

           3
    1      X  + 7

    2      X + 5

    3      X + K

    4      X + K
```

Let us examine the output. The numerical value of the

FACTORIZE call, the "4" that printed out, tells us that FA(4) is the last position in the array after the automatic re-dimensioning. So we print out FA(0) through FA(4).

FA(0) contains the largest constant, 75, that could be factored out of A. If there had been none, FA(0) would have contained the number 1. FACTORIZE doesn't factor the constant.

The factor $x^3 + 7$ was not further factored, since FACTORIZE can't generate the cube roots that would be needed. (It wouldn't even generate the SQRT(2) that would be needed if we wanted to factor $x^2 - 2$.)

The next factor is self-explanatory.

The next factor is repeated, because A is divisible by the square of X+K.

The reader who is curious about the process FACTORIZE goes through can eavesdrop on it by typing the command ON TRFAC$ before asking it to factor some expressions. Once you've seen enough such tracing, use OFF TRFAC$ to disable this extremely verbose mode. (The analogous command for tracing INT was presented earlier.)

2.5.1. Simple variable destinations

There is another form in which FACTORIZE can be used. If the second argument (more exactly, the value of the second argument) is not an array name, the factors will be assigned to a series of simple variables whose names are made up of the name of the second argument followed by the number of the factor. A message is printed to inform the user of the names of the variables set.

FACTORIZE

```
    FACTORIZE(A,W)$      % assuming W clear and not an array

         *** W0 W1 W2 W3 W4 ARE NOW NON-ZERO
                          % the message

    W2;                   % examine one of the factors

         X + 5
```

There is a still simpler form for using FACTORIZE. If we had omitted the second argument entirely, and had just written FACTORIZE(A)$ or FACTORIZE A$, the name FACTOR would have been assumed, and the message would have been

```
         *** FACTOR0 FACTOR1 FACTOR2 FACTOR3 FACTOR4
                                   ARE NOW NON-ZERO
```

If we have no need for storing the factors separately, but merely want to look at them, instead of FACTORIZE the ON FACTOR$ mode-changing command described in Chapter 3 should be used. ON FACTOR is not limited to polynomials: if the expression is a fraction, ON FACTOR will factor the numerator and also the denominator. (It uses the same computational code as FACTORIZE uses to do the actual factoring.)

2.6. The SOLVE function

SOLVE is "work in progress" being distributed with some REDUCE systems in its present incomplete state. Its goal is to enable solving numerous types of equations and systems of equations, exactly if possible, and otherwise, numerically (if there are no symbolic parameters) in the sense of determining intervals in which the solutions are certain to lie.

In some implementations of REDUCE, the command LOAD "SOLVE"$ must be given before the first use of SOLVE in a session.

In the present version of SOLVE, the solution(s) will be stored in a matrix called SOLN. If there are N equations in N unknowns, one solution will be stored in row 1 of the matrix: SOLN(1,1),

SOLN(1,2), ..., SOLN(1,N). If there is a second solution, those values of the N unknowns will be stored in row 2: SOLN(2,1), SOLN(2,2), ..., SOLN(2,N). If there are R solutions, the matrix SOLN will have R rows of N entries each.

Repeated solutions aren't repeated in the matrix. Instead, there is provided a second matrix, called MULTIPLICITY, with the same number R of rows but with only one entry in each row. The entry in any row of this matrix is just the integer that tells the multiplicity of the solution in the corresponding row of SOLN: so the entries are normally all equal to 1. (Negative integers may also be found in this array; this will be explained shortly.)

We spoke of solving equations, but more accurately we should have spoken of finding the values of the unknowns that make certain expressions equal to zero. That is, the equations should all be in the form

```
expression = 0,
```

and it will only be the expression(s), or variables equal to them, not the "= 0" part, that will be typed in. (This, too, may change in future versions of REDUCE.)

To solve

```
3*X + 2*Y - 36 = 0,
2*X -   Y - 17 = 0
```

we enter

```
SOLVE(LST(3*X + 2*Y - 36, 2*X - Y - 17), X,Y);
```

Note the symbol LST. The two expressions are separated by commas and enclosed in the parentheses following the "LST". Next are the two unknowns, X and Y.

The SOLN matrix prints out automatically. If this is not desired, enter the command OFF SOLVEWRITE$. (This OFF command accomplishes its purpose only if a SOLVE calculation has already been

The SOLVE function

made during the REDUCE session. If it hasn't, enter some simple SOLVE task, like `SOLVE(1,X)$`, before typing `OFF SOLVEWRITE$`.)

The output corresponding to the above example will be

```
SOLN(1,1)  :=  10

SOLN(1,2)  :=  3
```

1

The "1" is the "value" of the SOLVE function: it is the number, R, of solution rows in the SOLN matrix. Since it is 1 in this case, the MULTIPLICITY matrix will be a 1-by-1 matrix:

`MULTIPLICITY;`

```
MAT(1,1)  :=  1
```

Now let us try a singular but consistent system (the second expression is just double the first):

```
SOLVE(LST(3*X + 2*Y - 36, 6*X + 4*Y - 72), X,Y);

    SOLN(1,1)  :=  (2*( - ARBCOMPLEX(0) + 18))/3

    SOLN(1,2)  :=  ARBCOMPLEX(0)
```

1

The arbitrary constant ARBCOMPLEX(0) was generated. The "one" solution is in reality an infinite family of solutions, with any real or complex number a valid substitution for the arbitrary constant. (If later another such arbitrary constant will be needed, it will be called ARBCOMPLEX(1), and so on.)

Now let us shift our attention from systems of linear equations to single equations of more complicated types. Let us start with a quadratic equation (more exactly, a quadratic expression to be made equal to zero). Note that LST is not used when there is only one equation.

```
SOLVE(X**2 + 5*X + 100, X);

    SOLN(1,1) := ( - SQRT( - 375) - 5)/2

    SOLN(2,1) := (SQRT( - 375) - 5)/2
```

2

The matrix is now of different shape, to represent two different solutions for a single unknown.

Another example:

```
SOLVE(X**3 - 7, X);
                        (1/3)
    SOLN(1,1) := ( - 7      *(SQRT(3)*I + 1))/2
                    (1/3)
    SOLN(2,1) := (7      *(SQRT(3)*I - 1))/2
                (1/3)
    SOLN(3,1) := 7
```

3

The three complex cube roots of 7 are correctly indicated.

SOLVE solves cubic equations by Cardan's formula. Unfortunately this formula is nearly useless for cubics that have three real roots, because it gives the answers as expressions involving complex numbers, from which the imaginary part can be made to cancel out only with great difficulty. For example, if we ask SOLVE to solve the cubic whose roots are 2, 3, and 7, by typing

```
SOLVE((X-2)*(X-3)*(X-7), X);
```

we get, for one of the roots,

The SOLVE function

$$\text{SOLN}(1,1) := (\text{SQRT}(-3) *$$
$$(\text{SQRT}(-100) + 9*\text{SQRT}(3))^{(2/3)} -$$
$$(\text{SQRT}(-100) + 9*\text{SQRT}(3))^{(2/3)} +$$
$$8*\text{SQRT}(3)*(\text{SQRT}(-100) + 9*\text{SQRT}(3))^{(1/3)} -$$
$$7*\text{SQRT}(-3) - 7) \;/\; (2*\text{SQRT}(3) *$$
$$(\text{SQRT}(-100) + 9*\text{SQRT}(3))^{(1/3)})$$

We challenge the reader to determine whether this is equal to 2, or 3, or 7. (It depends, in fact, on which of the three possible complex cube roots each 1/3-power is assumed to represent.) This is discussed further as one of the Case Studies in Chapter 5, on page 278.

Let us test SOLVE on a fifth degree polynomial for which we know the answer:

```
A := (X-5)**3 * (X - AA) * (X + 2*BB)$

SOLVE(A, X);

        SOLN(1,1) := 5

        SOLN(2,1) := - 2*BB

        SOLN(3,1) := AA


        3

MULTIPLICITY;

        MAT(1,1) := 3

        MAT(2,1) := 1

        MAT(3,1) := 1
```

The MULTIPLICITY matrix indicates that the first solution arises from a three-fold factor of A.

(This was really easier for SOLVE than the cubic given just above. The multiple root 5 could be identified by a simple technique using Calculus, leaving only a quadratic which caused no difficulty.)

Now, an equation involving the SIN of the unknown:

```
SOLVE(SIN X - 2/3, X);

    SOLN(1,1) :=  - ASIN(2/3) + 2*ARBINT(2)*PI + PI

    SOLN(2,1) := ASIN(2/3) + 2*ARBINT(2)*PI


    2
```

This time, the arbitrary constant's name indicates that it must be chosen to be an integer. (ASIN is the principal branch of the arcsine function.)

Finally, let us give SOLVE a problem it can't do at all:

```
SOLVE(X**3 - SIN X, X);

                    3
    SOLN(1,1) := X   - SIN(X)


    1


MULTIPLICITY;

    MAT(1,1) := (-1)
```

The MULTIPLICITY matrix contains a negative entry, as a code to indicate that the corresponding entry in SOLN isn't a solution at all, but a factor of the given expression which SOLVE was unable to solve.

We have looked at examples of systems of linear equations,

The SOLVE function

which have single solutions, and at single equations of higher degree (or transcendental). What about systems of higher degree equations? This is beyond the capability of the present version of SOLVE. In other words, the SOLN matrix will either be 1-by-N, or R-by-1, at present.

Another aspect of SOLVE that is not implemented at this time is the numerical solution of equations. It is intended that if the command ON SOLVEINTERVAL$ has been given, real roots that can't be determined exactly will be reported as being in a range, in the form INTERVL(low,high).

2.7. LET and CLEAR

2.7.1. A first look at LET

LET can be thought of first as a version of the assignment operator for which the right-hand side is not evaluated.

Suppose that in the middle of a REDUCE session you decide that the variable XSQ should always stand for the square of X. The assignment

    XSQ:=X**2;

would accomplish this provided X is clear. But if X had a value at this time, say 5*W, XSQ would be assigned the value 25*W**2, and wouldn't reflect future changes in the value of X (only of W).

If instead you wrote

    LET XSQ=X**2;

the value assigned to XSQ would be X**2 itself, with the X not evaluated at this time. The X in the value of XSQ would be evaluated anew on each occasion that the value of XSQ is needed.

```
X:=5*W;
......
LET XSQ=X**2;
......
X:=ABC;
......
XSQ;
            ABC²
......
X:=PQR;
......
XSQ;
            PQR²
```

Exercise 2.7.1.

Try the above example as given (without the "....." lines, of course), and also with an assignment (:=) in place of the LET line for XSQ.

We repeat the warning that LET A=B does not work reasonably if B is a matrix variable or expression, and should be avoided. See page 56.

An important use of LET is in exploratory computation. Suppose some complicated expression is to be evaluated under a variety of circumstances, e.g. with various values assigned to some of the variables in it. After going several times through the cycle of changing the variables, typing the expression to be evaluated, changing the variables, typing the expression to be evaluated, ..., it should occur to the user to LET (say) S = the expression. Then whenever the expression is to be evaluated all he needs to do is to type S; ! This is even less work than the trick of using INPUT n which was described earlier, on page 22.

In reality what we have just described is a minor aspect of LET. Its major purpose is to enable the user to record a family of substitution rules, possibly of considerable complexity, which are to be obeyed automatically during computations. This is the subject of the following subsections. The collection of LET rules that are in effect at any particular time are sometimes referred to as the substitution

LET and CLEAR

environment.

There is a limit to the complexity of the substitution rules that can be stored. This limit varies from version to version of REDUCE, so we won't attempt to describe it here. If the rule is too involved, the error message

    *** SUBSTITUTION FOR ... NOT ALLOWED

will be issued, with the ... a copy of the left side of the rule, possibly in a cryptic form.

2.7.2. LET power = ...

It should suffice to give some annotated examples.

1.

    LET Y = A+B$

This causes any Y arising in a computation to be replaced by A+B; more precisely, by the result of evaluating A+B at the time the computation involving Y is performed, which is <u>not</u> necessarily the same as the result would have been of evaluating A+B at the time the LET rule was input.

This kind of LET statement, in which the "assignment" is to a variable, was discussed in the previous subsection, and is repeated here only for the sake of completeness. We point out that it is on the same logical level as an assignment made with the := operator. That is, an assignment Y := P+Q replaces a rule LET Y = A+B entered earlier, and vice versa. The only difference is that in the assignment case what is recorded in the substitution environment involves the current value of P+Q (which will not be P+Q if P or Q is not clear), whereas what is recorded for the LET rule is the expression A+B whether or not A and B are clear.

2.

   LET Y**3 = A+B$

   This causes Y**3 to be replaced by A+B in any expression in which it occurs during calculation, provided Y is clear at the time. (If Y is not clear, Y**3 continues to be the cube of the value of Y. However the LET rule is retained in REDUCE's memory and becomes effective when and if Y becomes clear.)

   Lower powers of Y are not affected. Higher powers of Y are replaced as would be expected: Y**4 by (A+B)*Y, Y**6 by (A+B)**2, Y**11 by (A+B)**3 * Y**2. Y to a symbolic power, like Y**Q with Q clear, is not affected.

   We now invite the reader to guess: with this LET rule in effect, what answer would REDUCE give to the computation of Y**4/Y**2? Would the division be done first, yielding the quotient Y**2 (to which the LET rule doesn't apply), or would the rule be applied to the numerator first, yielding (A*B)*Y/Y**2 which simplifies by cancellation to (A*B)/Y?

   The answer is less important (after all, the two are equal!) than the realization that the exact form of an answer may depend on the details of the way REDUCE works. In apparently similar computations, REDUCE may apply the various LET rules in the substitution environment and the various rules of algebraic simplification in different sequence, resulting in different-seeming answers. In Chapter 5, dealing with "concrete" applications of REDUCE, much of the effort will consist of coaxing REDUCE to apply rules in the order we want.

3.

   LET Y**3 = 0$

   This is a special case of the kind of rule just discussed, but should be mentioned separately because of its importance. With this rule in the substitution environment, powers of Y of Y**3 or above are automatically deleted from expressions. In an application in which

LET and CLEAR

Y represents a small quantity, this, which is usually referred to as "neglecting higher-order terms", may be appropriate in order to simplify expressions. This can greatly speed up computation, and frequently changes a problem that is "too big" into one that can be handled.

We give an example. If we need the coefficients of only a few of the lower-degree terms of a polynomial of very high degree, it is not necessary to waste computer time having COEFF expand an array to a huge size, and fill it, only to ignore most of the array. We can use a LET statement to delete, temporarily, the high-order terms. For example, if A is a polynomial in X and only the constant term and the coefficients of X and X**2 are required, we can use

ARRAY CO 2$

LET X3=0$**

COEFF (A,X,CO)$

CLEAR X3$**

This sets CO(0) through CO(2) to the desired coefficients. (The purpose of the CLEAR X**3 will be discussed on page 87.)

Exercise 2.7.2.

What would you expect

LET X3=0$**

X := A$

X3;**

to print? Guess first, then try it.

There is another way to tell REDUCE to treat powers of variables (and products of powers of variables) as zero. This uses the WEIGHT and WTLEVEL commands explained on page 101.

82   A Harder Look

2.7.3. MATCH power = ...

MATCH is a variant of LET that is of infrequent use (but see page 263). While `LET X**3=17` has an effect on all X**N with N equal to or greater than 3, `MATCH X**3=17` causes only X**3 to be replaced, and has no effect on X**4, X**5, etc.

Exercise 2.7.3.

What result would you expect to get from

`MATCH X**3 = AAA$`

`X**3 * (X**3 + 1);`

Guess first, then try it.

2.7.4. LET product = ...

1.

`LET X*Y = A+B$`

In any expression in which the product X*Y appears it will be replaced by A+B. If X**5 * Y appears, the rule changes it to (A+B) * X**4. If X**5 * Y**3 appears, we would get (A+B)**3 * X**2. If X or Y appears by itself, no change is made.

An example would be instructive.

`LET H*B=W$`

`Q:=A*B*C*D*E + A*B*C*D*E*F*G*H*I + C*D*H*I + Z;`

   `Q := A*B*C*D*E + A*C*D*E*F*G*I*W + C*D*H*I + Z`

REDUCE, knowing the rule LET H*B=W (among perhaps dozens of other rules), examines each term of every expression to see if it contains an H, and if so, whether it also contains a factor B. That being the case in the second term of Q, the factors H and B are deleted, and a factor W is inserted.

LET and CLEAR

Obviously when there are many LET rules in the substitution environment, REDUCE may slow down considerably!

2.

    LET H**3*B**4 = A+B$

With this rule in the substitution environment a substitution is made in any term containing H to at least the third power and B to at least the fourth power.

3.

    LET P*Q*R*S = A+B$

A rule can look for products of any number of factors.

2.7.5. LET sum = ...

This type of LET rule works quite differently from the LET product = ... form.

1.

    LET X+Y = W$

It is perfectly true that with this LET rule in the substitution environment any X+Y in any expression will be replaced by W, as the reader would expect. What the reader would not expect is that exactly one of the following will actually happen:

- Every X will be replaced by W - Y

- Every Y will be replaced by W - X

That X+Y is replaced by W is a consequence of this.

LET X+Y = W is just an alternative way of writing a certain one of the two rules LET X = W-Y or LET Y = W-X. But later we will see applications in which LET sum = ... is much more convenient than the simpler form of LET.

What determines whether it is the X or the Y that is replaced? (Obviously both can't be, for then we would have an endless loop.) There is no simple answer. It may change from one time you run REDUCE to the next, because the selection of the variable to be replaced depends on the internal state of REDUCE which in turn depends on what has been done earlier in that REDUCE run. The variables used in a REDUCE session have an internal ordering, which -- in most implementations -- is fixed for once and for all for the one-letter variables, and for other variables depends, sometimes in a highly irregular way, on the order in which they first appeared during the session. The variable that is replaced is the one that is the earliest in this ordering.

In many situations this uncertainty doesn't matter. For a dramatic example in which it does, see Exercise 2.7.14 on page 100.

(For some purposes the **KORDER** command, explained later, can override the internal ordering, but starting with REDUCE 3.2 **KORDER** doesn't affect which variable is singled out by the LET mechanism.)

2.

   LET A∗B + X∗Y = W$

Again, exactly what this does depends on the internal state of REDUCE. Either

- Every A∗B will be replaced by W - X∗Y, or
- Every X∗Y will be replaced by W - A∗B.

In either case, A∗B + X∗Y will of course become W.

Again, which part will be singled out for replacement depends on the internal ordering of the variables.

LET and CLEAR

2.7.6. LET operator = ...

If H, K, L are operators,

LET H(X) = X**10$

LET K(U,V) = U-V$

LET L(X)*A + L(1-X)*B = C$

will cause any occurrence of the symbol H(X) in any expression to be replaced by X**10; of K(U,V), by U-V; of L(X)*A by C - L(1-X)*B (or "vice versa", depending on the internal state of REDUCE).

If such a symbol H, K, or L appearing in such a rule being typed in was not previously declared to be an operator, the system will ask DECLARE ... OPERATOR?. Answer Y and strike RETURN.

It makes quite a difference whether X is clear or not when a LET H(X) = X**10$ command is entered. Let us look at an example.

X := ABC$

LET H(X) = X**10$

H(X);
$$10$$
$$ABC$$

This is what we would expect. But let us now give X a different value:

X := PQR$

H(X);

H(PQR)

Why didn't we get PQR^{10}? Because no rule for "H(X)" was in fact stored. H(X) is just H(X), which, evaluated at this time when X is PQR, becomes H(PQR). The rule which <u>was</u> stored was for "H(ABC)". Let's verify this:

```
H(ABC);
            10
         PQR
```

Let's compare this with what happens if an ordinary assignment ":=" is used in place of LET ... =.

```
X := ABC$

H(X) := X**10$

H(X);
            10
         ABC

X := PQR$

H(X);
         H(PQR)

H(ABC);
            10
         ABC
```

The only difference is in the response to the last input. The value stored was, as with LET, a value for "H(ABC)"; but the value was not X^{10}, but ABC^{10} (X being then ABC).

To summarize: If **LET** H(U) = V$ is entered at a time that U has the value UU and V has the value VV, the effect is as if **LET** H(UU) = V$ had been entered. If H(U) := V$ is entered at such a time, the effect is as if H(UU) := VV$ had been entered. The variable -- or, more generally, expression -- in the operator symbol's parentheses is immediately evaluated in both cases; the expression on the right-hand side is immediately evaluated only in the ":=" case.

At this point we note that if a number of LET declarations are to be made at the same time it is not necessary to type the word LET repeatedly. Instead of seven separate LET declarations one can enter

LET and CLEAR

```
LET X = Y**2,
    H(U,V) = U - V,
    COS(PI/3) = 1/2,
    A*B = C,
    L+M = N,
    W**3 = 2*Z - 3,
    Z**10 = 0$
```

Note the use of commas. The division into lines is of course only for convenience and legibility.

Reminder:

```
LET K(U,V) = U - V;
```

will cause K(U,V) to evaluate to U - V (assuming U, V are clear), but will not affect K(U,Z) or K(10,7) or K with any arguments other than precisely the two symbols U,V in that order (or expressions which evaluate to U,V). How to define more general rules is the subject of two of the subsections to come.

2.7.7. CLEAR

The user may delete any LET rule from the system by entering a CLEAR command of the same form (but without the "= ..."). If the LET rules

```
LET Z = A+B$
LET Y**3 = A+B$
LET X*Y = A+B$
LET P+Q = 999$
LET H(X) = X**10$
LET K(U,V) = U-V$
```

are all to be cancelled, enter

```
CLEAR Z$
CLEAR Y**3$
CLEAR X*Y$
CLEAR P+Q$
CLEAR H(X)$
CLEAR K(U,V)$
```

As with LET, these can be combined if desired:

```
CLEAR Z, Y**3, X*Y, P+Q, H(X), K(U,V)$
```

The more general types of LET declarations about to be described can also be cancelled by using CLEAR.

2.7.8. FOR ALL X LET ... = ...

If a substitution for all possible values of a given variable in an expression is required, the declaration FOR ALL (or FORALL) is used.

```
FOR ALL X LET K(X,A) = X**A$

FOR ALL X,Y LET H(X,Y) = X-Y$

FOR ALL X LET X**3 = O$

FOR ALL P LET H(X**P) = P$
```

The first of these declarations would cause K(Q,A) to be evaluated as Q**A. K(Q,B) would be left unchanged (unless B has the value A). If the operator symbol K is used with more or fewer argument places, not two as in the LET statement, the LET would have no effect.

The second of these declarations would cause H(A,B) to be evaluated as A-B, H(U+V,U+W) to be V-W, etc.

The third would cause all cubes and higher powers in any expression to be replaced by zero (but see the exercises below).

The fourth would replace H(X**5) by 5, H(X**PQR) by PQR, but wouldn't affect H(Y**5) (different variable), H(X**1) (treated as H(X), no exponent present), H(X**(-5)) (normally treated as H(1/X**5), a fraction rather than a power; but see OFF MCD -- "OFF Make Common Denominator" -- in Chapter 3).

Where we used X, Y, and P in the examples, any variables

LET and CLEAR

could have been used. This use of a variable doesn't affect the value it may have outside the LET statement:

    X := ABC$

    FOR ALL X LET K(X,A) = X**A$

    K(PQR,A);

                A
             PQR

    X;

             ABC

However, you should remember what variables you actually used. If you want to delete the rule subsequently, you must use a CLEAR command with the same variables. For example, to cancel

    FOR ALL P LET H(X**P) = P$

you must use

    FOR ALL P CLEAR H(X**P)$

and not

    FOR ALL Q CLEAR H(X**Q)$

Exercise 2.7.4.

What output would you expect from

    FOR ALL X LET X**3 = 0$

    (A+B)**3;

Guess first, then try it.

Exercise 2.7.5.

Try the same exercise, but first enter the command

    OFF EXP$

In the OFF EXP mode expressions aren't always expanded. See Chapter 3. (When done with this exercise, enter ON EXP$ to return REDUCE to its normal mode.)

We want to point out a difference between the statements

    FOR X := 1:100 DO H(X) := ...

and

    FOR ALL X LET H(X) = ...

The first stores 100 individual values for $H(1)$, $H(2)$, ..., $H(100)$. The second stores a single rule which is examined and obeyed whenever an expression using H as a one-place operator is being evaluated and simplified.

Exercise 2.7.6.

Assume that H is an operator and that assignments have been made, using := or LET, to $H(0)$, $H(1)$, ..., $H(100)$, and to no other H( )'s. Does

    FOR ALL I CLEAR H(I)$

clear all of them? If not, find a statement that does.

Exercise 2.7.7.

Will the same apply if H is not an operator but an array dimensioned to 100? Is there a simple way to set all of $H(0)$ to $H(100)$ to zero?

We now give a lengthy annotated example.

LET and CLEAR

```
FOR ALL X,Y,Z LET F(X + Y*Z) = F1(X) + F2(Y) + F2(Z)$

F(A + B*C);

        F1(A) + F2(B) + F2(C)       % as expected

F(A + C + C);

        F1(A) + F2(C) + F2(2)       % 2 taken as a factor

F(A + 1*C);

        F(A + C)                    % 1*C not a product

F(A + B*B);

                2                        2
        F(A + B )                   % B  not a product

F(A - B*C);

        F(A - B*C)                  % not accepted by rule

F(B*C - A);

        F1( - A) + F2(B) + F2(C)
                    % accepted with -A matching the X

F(C1*C2*C3 + C4*C5*C6);

        F1(C1*C2*C3) + F2(C4*C5) + F2(C6)
                    % one product taken as X,
                    % other product partially split
```

One could argue that F(A - B*C) should have been accepted by parsing it as, perhaps, F(A + (-B)*C). But while REDUCE's pattern matching ability is versatile, it is not infinitely so!

Exercise 2.7.8.

Add more LET rules to the one rule studied in the example above in order to take care of the degenerate cases in which one might expect it would work but it does not. Test the effectiveness of the resulting collection of LET rules.

For reasons of efficiency a cutoff was programmed into the

pattern search mechanism of REDUCE. The following example shows a case in which REDUCE didn't find a pattern that was actually there:

```
OPERATOR Z$

A:=Z(Q1*Q2*Q3*Q4*Q5*Q0);

        A := Z(Q1*Q2*Q3*Q4*Q5*Q0)

FOR ALL X LET Z(X*Q0) = X$

A;

        Z(Q1*Q2*Q3*Q4*Q5*Q0)
```

We should have gotten `Q1*Q2*Q3*Q4*Q5` as the answer, but didn't. The reason is that REDUCE gave up looking for a Q0 factor before it got to it, there being five factors in front of it. If we had only four factors in front, REDUCE would have found the Q0 and applied the LET rule:

```
Q3:=1$

A;

        Q1*Q2*Q4*Q5
```

There is a special use of FOR ALL X LET ... in connection with the symbol WS, Workspace. Recall that WS always stands for the result of the last "ordinary" statement executed. If a result like

$$A*X**5 + B*X**2 + C$$

was obtained from some operations, and the user decides that this would be a useful expression to save as a general rule F( ) where F is to be an operator, it is not necessary to retype it:

```
FOR ALL X LET F X = A*X**5 + B*X**2 + C$
```

Instead, one can just type

```
FOR ALL X LET F X = WS$
```

or

LET and CLEAR

        FOR ALL X SAVEAS F X$

quite analogously to the simple uses of WS and SAVEAS described in Chapter 1. Then F(R), for example, would be A*R**5 + B*R**2 + C.

If REDUCE prints an error message pertaining to a FOR ALL ... LET statement, in which some or all of the FOR ALL statement is reproduced, you may observe that any variable that was listed in the FOR ALL part will have its name preceded by an equal sign: X, for example, will appear as =X . This is how REDUCE reminds itself that X here is a general (or "universally quantified") variable.

We close this subsection with a "practical" application of operators and the FOR ALL ... LET statement. Suppose we have an expression W which is the sum of a large number of terms, and we need the sum of the squares of the terms. At first glance this has nothing to do with operators. But the first problem is to somehow isolate the individual terms from each other:

    OPERATOR H$
    WW := H(W)$ % create expression H(...);

    FOR ALL P,Q LET H(P+Q) = H(P) + H(Q)$
    WW := WW$ % create H( ) + ... + H( );
    FOR ALL P,Q CLEAR H(P,Q)$

For example, if W had been A+2*B+3*C we would now have, for WW,

        H(A) + H(2*B) + H(3*C).

Now to replace the H's by squares:

    FOR ALL P LET H(P)=P**2$
    WW := WW;   % create ( )**2 + ... + ( )**2;
    FOR ALL P CLEAR H(P)$

With the value of W given above, the WW obtained is

            WW := A^2 + 4*B^2 + 9*C^2

which is what was desired.

Exercise 2.7.9.

Try the above. What would you expect to get if W were a sum and difference of terms? Why doesn't it work if W is $A/2 + B/3$? Can you revise the method to work for this, too?

This example can be used to illustrate the difficulties REDUCE can have with really large problems. If you try this with many more terms, you will discover that both the time and the working space REDUCE needs to do the above are roughly proportional not to the number of terms in W, as might be expected, but to the square of the number of terms. The culprit is the step in which the H( ) operator is distributed over the additions. There is a much more efficient way to do this than to use the FOR ALL ... statement of REDUCE -- but it requires programming in RLISP, which is beyond the scope of this book!

2.7.9. FOR ALL ... SUCH THAT ... LET ... = ...

If a substitution is desired for more than a single value of a variable in an operator or other expression, but not all values, a conditional form of the FOR ALL ... LET declaration can be used. For example,

    FOR ALL X SUCH THAT X<0 LET H(X)=0;

will cause H(-5) to be evaluated as 0, but H of a positive number would not be affected.

This example leaves something to be desired. We would get an error message if we subsequently attempted to evaluate an expression containing, say, H(A) with A clear! This is because the < operation can only be used between numbers, and the clear A isn't a number. To remedy the situtation:

    FOR ALL X SUCH THAT NUMBERP X AND X<0 LET H(X)=0;

will function as before if X is an number, but leaves H( ) in any

LET and CLEAR

expression unchanged if the ( ) doesn't contain a number or number-valued expression. Here we have used the Boolean function NUMBERP mentioned in Chapter 1.

These LET rules, like all others, can be cancelled by using CLEAR. Simply repeat the LET rule to be deleted, using CLEAR in place of LET, and omitting the equal sign and right-hand part. The same dummy variables must be used in the FOR ALL part, and the boolean expression in the SUCH THAT part must be written the same way. (Placing of blanks doesn't have to be identical.) For example:

    FOR ALL X SUCH THAT NUMBERP X AND X<0 LET H(X)=0$

can be erased by the command

    FOR ALL X SUCH THAT NUMBERP X AND X<0 CLEAR H(X)$

Exercise 2.7.10.

First, note the output resulting from the commands

    OPERATOR H;

    H(5) := A$

    FOR ALL X LET H(X) = B$

    H(5);

If the version of REDUCE you are using gives the answer A, there is nothing for you to do. But in some versions, the LET rule is attended to before the actual stored value of H(5) would be noted, and you get B as the answer. CLEAR the LET rule, and replace it by a rule that allows H(5) to be evaluated as A while causing <u>every other</u> use of H as a one-place operator to be evaluated as B.

Suppose we want B to be a symmetric two-place operator, so that B(5,9) and B(9,5), for example, are considered equal. Assume B will always be used with two numbers in the operator places. We can then standardize B(I,J) by insisting that I be never greater than J:

```
FOR ALL I,J SUCH THAT I>J
    LET B(I,J) = B(J,I)$
```

With this LET rule in effect, if B(9,5) ever arises it would automatically be changed to B(5,9).

Exercise 2.7.11.

Write two LET rules which, together, would make B be a symmetric three-place operator, with a form like B(9,1,3) automatically changed to B(1,3,9), and B(5,5,5) left alone. Assume all three operator places will always be occupied by numbers. We want B(I,J,K) to be transformed to a standard form in which $I \leq J \leq K$.

Note: Actually these examples and exercises on making an operator symmetric are <u>only</u> examples and exercises. If the practical REDUCE user wants the operator B to be symmetric he only needs to declare it to be:

SYMMETRIC B$

This will do all we asked, except that the standard ordering of numerical values in the operator places happens to be the decreasing order: B(9,7,3). It will also do something we wouldn't be able to program ourselves without going into RLISP: symbolic argument places would be put into a standard order too. Thus B(VV,UU) would be changed to B(UU,VV) (or vice versa -- there is no way to predict which).

Exercise 2.7.12.

Write LET rules which, together, would make B be an anti-symmetric (skew-symmetric) three-place operator. What this means is that a form like B(9,7,3) should be automatically changed to -B(3,7,9) [minus sign, because a single interchange 9 <---> 3 turns one into the other]; B(9,3,7) = B(3,7,9) [because it takes two interchanges (an even number) to go from one to the other]; and B(5,8,5) = 0 [because two places are equal].

You may assume all three places will always contain numbers.

LET and CLEAR

This can be done with four LET statements, which can even be condensed into two.

Note: Again, the practical REDUCE user doesn't need to make an operator anti-symmetric this way! The declaration

    ANTISYMMETRIC B$

would do all this. Again, the standard ordering for numbers is the decreasing order, and again, the antisymmetry rule is obeyed for symbolic operator places too.

2.7.10. LET DF(...) =

LET can be used for the introduction of rules for the differentiation of user-defined operators.

    FOR ALL X LET DF(TAN X,X)=(SEC X)**2$

Notice that a FOR ALL X prefix is necessary. (This particular rule, that the derivative of the tangent is the square of the secant, may be already predefined in some versions of REDUCE.)

If no such rule is supplied, a calculation leading to DF(TAN X,X) or DF(TAN Y,Y) would have these symbols, like DF(TAN X,X), appear unchanged in the answer.

If an operator will be used with several positions filled, partial derivatives can be declared similarly.

    FOR ALL X,Y LET DF(F(X,Y),X)=2*F(X,Y)$

    FOR ALL X,Y LET DF(F(X,Y),Y)=X*G(X,Y)$

Rules may be supplied for operators with any number of arguments. Notice that all the dummy arguments of the operator (here X and Y) must be listed in the FOR ALL part of the command. Mysterious behavior may be encountered if this is not done.

If the differentiation rule for some argument position is not given, the DF operation will return as result an expression in terms of DF. For example, if the rule for the differentiation with respect to the second argument of F is not supplied, the result of evaluating of an expression containing $DF(F(X,Z),Z)$ would contain the symbol DF(F(X,Z),Z) unchanged.

2.7.11. Overlapping rules

CLEAR is not the only way to delete a LET rule. A new LET rule, identical to the first except for a different expression after the equal sign, replaces the first.

A new LET rule closely related to the first may or may not replace the first. In fact there are quite a number of possibilities:

- The new rule may completely replace the old rule. If the new rule is cleared, subsequently neither rule is in the system.

- The new rule may override the old rule. Only the new rule is obeyed, but if it is cleared the old rule stands revealed and functional.

- The new rule may be stored but ignored, the old rule still being in control. If the old rule is cleared, the new rule becomes effective.

- If the rules apply under different but overlapping conditions, both rules could be stored but are effective under different circumstances.

In the current version of REDUCE a rule LET X**10 = ... completely replaces a rule LET X**5 = ... :

LET and CLEAR

```
LET X**5 = A$

X**7;
           2
        A*X

LET X**10 = B$

X**13;
           3
        B*X

X**7;
        7
       X
```

(The LET X**10 rule does what it is supposed to do, but the LET X**5 rule no longer functions. Is it still there, hidden "under" the LET X**10 rule? No:)

```
CLEAR X**10$

X**7;
        7
       X
```

Of course the above applies only to LET rules concerning powers of the same variable. A LET X**5 rule and a LET Y**10 rule co-exist freely in the system, and each performs its job when appropriate.

Exercise 2.7.13.

Discuss as many other ways as you can think of that a LET X**5 and a LET X**10 rule might interact in a REDUCE-like system.

In general the user should try to avoid having in effect several LET rules pertaining to the same expression. No guarantee can be given as to which rules will be applied by REDUCE or in what order.

It is best to CLEAR an old rule before entering a new related LET rule.

Exercise 2.7.14.

Suppose we have the three consecutive lines

```
LET AA + BB = 0$
LET AA + CC = 0$
AA + BB;
```

Will the answer to the last line be zero? <u>Hint</u>: It depends on the relative positions of AA, BB, CC in the default ordering of variables. Try it as shown; then CLEAR the two rules and input the three lines using the symbols AA, BB, CC permuted in some way.

The best intentions won't always prevent relevant rules from being applied in an unexpected order. Observe the following example:

```
OPERATOR F$
FOR ALL X,Y LET F(X+Y) = PPP$
FOR ALL X,Y LET F(X-Y) = MMM$
F(A+B);
        PPP
F(A-B);
        MMM
```

No surprise here. But compare with

LET and CLEAR

```
OPERATOR G$

FOR ALL X,Y LET G(X-Y) = MMM$

FOR ALL X,Y LET G(X+Y) = PPP$

G(A+B);

        PPP

G(A-B);

        PPP
```

Why didn't we get MMM?

The answer lies with the way REDUCE represents A-B, namely as A+(-B). For two LET rules of the same general kind the most recent one is checked first. $G(A+(-B))$ is an instance of $G(X+Y)$ -- namely with X=A and Y=(-B) -- so the first test made, for a $G(X+Y)$ form, responded positively. For the operator F, F(A-B) was accepted by the first test made, for F(X-Y); F(A+B) wasn't accepted by the first test, only by the second, because the first test was looking for a minus sign as well as an addition.

Exercise 2.7.15.

Suppose an operator F is supposed to have the properties

```
FOR ALL X,Y LET F(X+Y) = F(X) + F(Y)$

FOR ALL X,Y LET F(X-Y) = F(X) - F(Y)$

FOR ALL Y LET F(-Y) = - F(Y)$
```

Are all three of these LET rules necessary? Test your answer.

2.7.12. WEIGHT

The commands LET X**5 = 0 and LET Y**5 = 0 tell REDUCE that X**5 and Y**5 are each negligible, but would have no effect on the product X**3 * Y**2 even though logically that product ought to be considered negligible too. (The two LET statements

imply that X and Y are of the same order of magnitude, so
$X**3 * Y**2$ would be of the same order of magnitude as $X**5$ or
$Y**5$.)

The pair of commands

    WEIGHT X=1,Y=1$

    WTLEVEL 4$

(written instead of the LET rules) accomplishes what we want. It says that if a product contains a power of X, or a power of Y, or both, and the sum of the exponents of X and Y is greater than 4, then the product should be replaced by zero. (If X or Y is missing, count 0 for its exponent.)

Notice that the WTLEVEL to be specified is one less than the power 5 in the LET statements, because in the LET case exponents "greater than or equal" to 5 were what mattered, while with WTLEVEL the relevant relation is "greater than".

This example illustrates the WTLEVEL command, but doesn't shed light on the WEIGHT command. Suppose instead we were trying to improve on

    LET X**5=0,
        Y**10=0$.

We would want any of $X**5$, $X**4 * Y**2$, ..., $X * Y**8$, $Y**10$ to be replaced by zero. The appropriate pair of commands is

    WEIGHT X=2,Y=1$

    WTLEVEL 9$.

This says that a product is replaced by zero if 2 times the exponent of X, plus 1 times the exponent of Y, is greater than or equal to 10 (so exceeds 9).

LET and CLEAR

Exercise 2.7.16.

Write and test WEIGHT and WTLEVEL commands to replace

    LET X**3=0,Y**4=0,Z**5=0$.

Hint: The least common multiple of 3,4,5 is 60, so what WTLEVEL should be used?

WEIGHT commands are cumulative. If a series of WEIGHT commands is given, like

    WEIGHT X=7$

    WEIGHT Y=9$

    WEIGHT A=3$

    WEIGHT X=5$

the weights of X,Y,A are 5,9,3. (The first weight assignment to X is replaced by the second.)

The simple CLEAR command, such as **CLEAR X$**, has among its many effects the deletion of any weight assignment X may have.

The use of a variable is restricted while it has weight. If X has weight, DF(...,X), COEFF(...,X,...) and several other function calculations give an error message:

            ***** X invalid as kernel

A new WTLEVEL command replaces the level set by a former one. If no WTLEVEL was ever given during the REDUCE run, a WTLEVEL 1 is assumed.

Exercise 2.7.17.

Explain what WTLEVEL 1 really means. Consider variables of weight 1 and also variables of higher weight.

2.7.13. Complex assignments

In most programming languages an assignment like

    A+B := C;

would be an error that only beginners would be likely to make. In REDUCE the assignment operator ":=" has a special LET-like behavior if the left side is not a variable but a more complicated expression. If A, B, C are clear the assignment has exactly the same effect as

    LET A+B = C;

If any of A, B, C are not clear the ":=" assignment has the same effect as a LET statement composed of A, B, and the value of C. For example, if the value of A is AA, of B is BB, and of C is CC, then

    A+B := C;

has the same effect as

    LET A+B = CC;

(Recall that, by contrast, LET A+B = C does not concern itself with the values, if any, of any of A, B, and C.)

We have been dwelling on A+B := ... assignments; the analogous behavior obtains with A*B := ... or any other expression on the left side.

2.8. WRITE

In simple cases no explicit output commands are necessary in REDUCE. The value of any expression is automatically printed out if a semicolon is used as a delimiter (unless it's inside a structure like << ... >>). There are, however, several situations in which explicit WRITE commands are useful.

WRITE

- We may wish to output something each time the statement within a loop construct like FOR, WHILE, or REPEAT is repeated.

- We may wish to enable a procedure to output intermediate results or status information while it is running.

- We may wish to label our results in special ways.

These and similar situations require the use of the WRITE command. This consists of the word WRITE followed by the item to be printed, and followed by a terminator. It makes no difference which terminator is used. There are three kinds of items that can be used:

- Expressions (including variables and constants). The expression is evaluated, and the result is printed out.

    WRITE B/C;   or   **WRITE B/C$**

  Note that no parentheses are required.

- Assignments. The assignment is performed, and both sides of the assignment are printed out (the destination variable, a ":=" sign, and the value which was just now saved there). The form is almost exactly the same as the way an assignment followed by a semicolon prints out normally. ("Almost exactly": the difference will be explained shortly.)

    WRITE S:=B/C;

- Arbitrary strings of characters, preceded and followed by double-quote (") marks. Strings print exactly as quoted.

    WRITE "END OF CASE 5";

A WRITE can also be used to print several items, by listing the items one after the other with commas between. The items specified by a single WRITE statement print side by side, with no automatic space between, on one line. To leave a space between two items, include a quoted space ...," ",... in the list of items for WRITE to print.

To print two lines, use two separate WRITEs. They will print "double spaced", or more widely apart if exponents are present. (A single line which is too long is broken automatically, but not necessarily at a logical place.)

Examples:

- If M is an array, Q=3, A=X+5, B is clear, and C=123, then

    WRITE M(Q):=A," ",B/C," THANK YOU";

  will set M(3) to X+5 and print

    M(Q) := X + 5 B/123 THANK YOU

  The only reason there are blanks between the 5 and the B, and the 3 and the T, are the blanks in the quoted strings (the " " items in the WRITE statement).

- To print a table of the squares of the integers from 1 to 20:

    FOR I:=1:20 DO WRITE I,"   ",I**2;

  Between the two quotation marks we used a single keystroke, the TAB key, to prepare this statement. Unfortunately this will not work on all systems. Some have no TAB key, or the TAB key has a special function or does nothing. But in many systems, TAB will cause a jump of the printer or display to a built-in point to start the second column. Caution: Don't use TAB in WRITE statements if the answers to be printed out may have exponents. The exponents may not print in their proper places. In this example we are only printing numbers (the values of I**2), so there will be no trouble.

- To print a table of the squares of the integers from 1 to

20, and at the same time store them in positions 1 to 20 of an array A:

```
FOR I:=1:20 DO <<A(I):=I**2; WRITE I," ",A(I)>>;
```

- The reader would expect that if we had used

```
FOR I:=1:20 DO WRITE A(I):=I**2;
```

we would get a useless repetition of an unchanging "A(I) := " -- with different right-hand sides -- down the page. Here is where the difference occurs at which we hinted before. The sequence of lines

```
J:=5$

A(J):=J**2;
```

would indeed print A(J) := 25. (Quick -- why did we use J in this illustration instead of I?). But when we have a WRITE in a FOR statement and the left-hand side of the assignment is an array place or operator "place" containing the variable of the FOR iteration, then the value of that variable is inserted in the printout:

```
FOR I:=5:6 DO WRITE A(I):=I**2;

    A(5) := 25

    A(6) := 36
```

Exercise 2.8.1.

Use a WRITE command to print the values of two simple expressions, of your own choice, side by side. What must be done to make the output more readable?

2.9. Grouping

The grouping symbols << ... >> were introduced in the Overview chapter in connection with FOR ... DO, WHILE ... DO, and REPEAT ... UNTIL, as a way to exercise control over the repetition of a group of statements collectively. Here these uses of grouping will be reviewed, and other uses given.

2.9.1. Grouped statements

Situations frequently arise in which REDUCE expects only a single statement to be written, but the application demands more than one. We saw examples when we discussed the FOR ... DO, WHILE ... DO, and REPEAT ... UNTIL constructions in the Overview chapter. And there we saw the solution: separate the statements by semicolons (or, equivalently, dollar signs) and enclose the resulting list of statements within the symbols << and >>. Then when -- or if -- it's time for REDUCE to execute the group << ... >>, the enclosed statements are executed one after another.

In this context the words BEGIN ... END can be used in place of the << ... >> symbols.

In the next subsection, which deals with the use of a group as an expression rather than as a statement, the reader will be warned against writing a semicolon (or dollar sign) after the last item in the group, that is, just before the closing >> symbol. For a group used as a statement, no harm arises from including an unecessary final terminator.

One use of grouped statements is to gain some efficiency in the interactive dialog with REDUCE. Suppose you want REDUCE to carry out several time-consuming computations one after the other, and don't want to wait for each computation to finish before typing in the command to carry out the next one. You can type in a single grouped statement encompassing all the commands:

   <<A:=..., B:=..., C:=...,   ..., Z:=...;>>;

Grouping 109

2.9.2. Groups as expressions

In REDUCE the distinction between statements and expressions is one of emphasis rather than one of syntax. A statement is a program part that causes some action (for example, an assignment, or a WRITE) to take place; an expression is a program part whose <u>value</u> is of interest (for example, for use as part of an assignment statement, or to be printed out.) But statements of most types also have a meaningful value. For example, assignments can be printed out, and can themselves be used in expressions such as in

    A := P + (B:= Q + R);

which first sets B to Q+R and then sets A to P+Q+R, Q+R being the value of the inner assignment statement.

And the evaluation of an expression frequently causes actions to take place which we could consider to be statement-like.

If a group << ... >> is used as an expression, its value is the value of the last item in the group. For example:

    <<A:=B+C; P:=A+D>>

is an expression whose value is the newly acquired value of P, namely B+C+D. In the course of its evaluation A is set to B+C and P to B+C+D.

Expressions can be used in many ways. For example, if we typed in that group by itself, followed by a semicolon:

    <<A:=B+C; P:=A+D>>;

the value of the group, i.e. the new value of P, will print out. (Notice that the semicolon is <u>after</u> the >> symbol!) The value of A, while calculated, is NOT printed out, even though its calculation is followed by a semicolon, because only the value of the group is printed automatically, not the values of the intermediate expressions. (To print an intermediate expression, insert a WRITE).

If we wanted the value of the group to be the newly acquired value of A, instead, we could write

    <<A:=B+C; P:=A+D; A>>

The last item in the list is A. Its value B+C is taken as the value of the entire group, but meanwhile the assignments to A and P have been made.

Let us use the group construction in a FOR ... SUM to calculate the sum of the first five terms of the power series for e^X. In the following, each term is called U in turn, and each U is calculated from its predecessor. (U is initialized to 1.) In addition to calculating the sum, each term used in the sum is printed out.

```
U:=1$

S:=FOR I:=1:5 SUM
        <<OLDU:=U;
          WRITE U;
          U:=U*X/(I+1);
          OLDU>>;

       1

       X/2

        2
       X /6

        3
       X /24

        4
       X /120

              4         3        2
       S := (X   + 5*X   + 20*X   + 60*X + 120)/120
```

Note that are variable OLDU had to be introduced, since U was already replaced by the next U before the old U was needed for the summing.

As another example, suppose we plan to introduce the LET rule

    FOR ALL X LET X**5 = 0$

Grouping   111

and want to know every time it is used.

    Q := A**5 + A**6 + B**3 + B**6 + C**20;

             6   5   6   3   20
        Q := A + A + B + B + C

    FOR ALL X LET X**5 = <<WRITE "REDUCED ",X; 0>>$

    Q;

        REDUCED A

        REDUCED A

        REDUCED B

        REDUCED C

         3
        B

The value of the << ... >> here is always zero, as required, but a WRITE statement is executed every time the zero is "evaluated".

Caution: Do not put a semicolon (or dollar sign) after the last item in the << ... >> if it is to be used as an expression! If there is a terminator just before the closing double angle bracket: ;>> then there is technically a "dummy statement" (also called "empty statement") at the end of the group. The "value" of the dummy statement is zero (more technically, NIL which in most contexts is treated as zero by REDUCE). So including a terminator at the end of the group causes the value of the group to be zero.

If a group is to be used as an expression, BEGIN ... END is not equivalent to the << ... >> symbol pair. As an expression BEGIN ... END has the value zero (or NIL), unless the word RETURN is placed in front of the last item.

We rewrite some of the examples from this subsection in this notation:

```
BEGIN A:=B+C; P:=A+D; RETURN A END;

T:=1$
S:=FOR I:=1:5 SUM
        BEGIN OLDU:=U;
            WRITE U;
            U:=U*X/(I+1);
            RETURN OLDU
        END;

FOR ALL X LET X**5 = BEGIN WRITE "REDUCED ",X END$
```

Observe that in the last example we have no RETURN 0, since by omitting the RETURN entirely we get the value zero anyway.

More on this topic when we study Procedures.

The grouping symbols << >> can be considered to be superparentheses. Suppose we have a LET statement

```
LET A*B=5$
```

in the substitution environment. Observe the calculation

```
A*C/(A*B);

        C/B
```

The A's were cancelled before the LET rule could be applied to the denominator. On the other hand,

```
A*C/<<A*B>>;

        (A*C)/5
```

The << >> symbols forced the denominator to be calculated and simplified completely before its relationship with the numerator was considered.

Exercise 2.9.1.

(Compare with Exercise 2.7.3 on page 82.) What result would you expect to get from

Grouping

```
MATCH X**3 = AAA$

<<X**3>> * (X**3 + 1);
```

Guess first, then try it.

2.10. IF ... THEN

Constructions involving IF ... THEN are used both for statements and for expressions in REDUCE. While the line between the two isn't sharp, as we have already seen, we will treat them separately below.

2.10.1. IF ... THEN statements

Whether an action should be taken, or not, can be made to depend on the value of a Boolean expression, i.e. a test.

```
IF X=Y THEN A:=B+C$
```

The meaning of this should be obvious:

- If X and Y are equal (when each has been evaluated and simplified as much as possible) then A is set to the value of B+C.

- If X and Y are not equal, no action is taken (and in particular A remains unchanged).

Most users will have little occasion to enter such a command by hand. If it matters whether or not X and Y are equal, the user could first find out by seeing whether or not the command "X-Y;" prints zero or not, and accordingly type in or not type in the command A:=B+C$. (One exception might be the case that the user knows that X-Y is either zero or a very lengthy expression whose printing would take an inordinate amount of time.)

The most common use for IF ... THEN statements is inside procedures, and we will defer giving meaningful examples until the chapter on that topic. We will merely note here that the controlled statement (the statement following the "THEN") is more often than not a grouped statement << ... >>, not just a simple assignment as in the illustration.

Also available is the IF ... THEN ... ELSE statement form:

    IF X=Y THEN A:=B+C ELSE D:=E+F$

Here the statement `A:=B+C` is executed if X and Y are equal, and `D:=E+F` is executed if they are not.

More often than not, grouped statements will be desired after the THEN and after the ELSE:

    IF X=Y THEN << ... >> ELSE << ... >>$

Sometimes we make some test -- i.e., evaluate a Boolean expression -- at one stage of a computation, and require that the path the computation takes at some later time depend on the result of the test. Unfortunately, the value of a Boolean expression can not be saved directly as the value of a variable in REDUCE. That is, REDUCE doesn't support Boolean variables. If the value of a Boolean expression must be remembered, one can set a variable equal to, say, 1 if it is true, 0 if it is false (or any other convenient coding):

    IF NUMBERP X AND NUMBERP Y AND X<Y THEN RR:=1
                                      ELSE RR:=0 $

Then the numeric value of RR can be tested later:

    IF RR=1 THEN ... ELSE ... $

2.10.2. IF ... THEN expressions

We begin with an example.

    A:=IF X=Y THEN B+C ELSE B-C$

On the right hand side of the ":=" there appears an expression which

IF ... THEN

is equal to B+C or B-C according to the outcome of the X=Y test. Instead of using an IF ... THEN ... ELSE expression we could have in this case used an IF ... THEN ... ELSE statement:

```
IF X=Y THEN A := B+C ELSE A := B-C$
```

But there are many examples in which this alternative is not practical. Suppose for example, that we want to add

$$\frac{1}{1-N} + \frac{1}{2-N} + \frac{1}{3-N} + \frac{1}{4-N} + \frac{1}{5-N} + \frac{1}{6-N} + \frac{1}{7-N}$$

omitting the term (if any) in which the denominator is zero. We can write:

```
X:=FOR I:=1:7 SUM
        IF I=N THEN 0
              ELSE 1/(I-N)$
```

When this assignment is made, if N has an assigned value and it is one of the numbers 1,2,...,7, the corresponding term is left out (by summing a zero in its place). Of course if N is clear when the sum is formed, and is <u>subsequently</u> set to one of these numbers, X will have a zero denominator:

```
CLEAR N$

X:=FOR I:=1:7 SUM
        IF I=N THEN 0
              ELSE 1/(I-N)$

N:=5$

X;

      ***** ZERO DENOMINATOR
```

If we set X with a LET instead of := the result would be different, since each time X is evaluated the then-current N would be used in the test:

```
LET X = FOR I:=1:7 SUM
          IF I=N THEN 0
                  ELSE 1/(I-N)$

N:=5$

X;

        ( - 7)/12
```

The skeptic who wants to see if the IF really works can put a tattle-tale WRITE into the expression:

```
LET X = FOR I:=1:7 SUM
          <<TERM := IF I=N THEN 0
                            ELSE 1/(I-N);
            WRITE I," ",TERM;
            TERM>>$

CLEAR N$

X;

N:=5$

X;
```

We leave the running of this to the reader.

An IF ... THEN used as an <u>expression</u> should always have an ELSE ... part, since the expression must have some value in either case. Actually the value is taken as zero if the condition is false and the ELSE part is omitted, but dependence on this is considered bad programming style. One should write an explicit ELSE 0 if that is what is intended.

Exercise 2.10.1.

Write a single expression whose value is 1 if the expression A doesn't contain the variable X, and 0 if it does. (Compare Exercise 2.1.3.)

2.11. PART and setting a PART

2.11.1. PART

NUM and DEN allow one to decompose a quotient into its numerator and denominator. PART allows any expression to be decomposed in any desired way.

To use PART, it is necessary to know what the expression looks like when printed out, because the order of the Parts corresponds to the order of the parts of the printout. Let us consider an example.

    EE:=B*C*D*A;

        EE := A*B*C*D

Thus EE has four Parts: A, B, C, and D. The PART function will pick out any one of them:

    F:=PART(EE,3);

        F := C

As a more complicated example:

    EE:=(A+B)*(C+D);

        EE := A*C + A*D + B*C + B*D

    F:=PART(EE,2);

        F := A*D

    G:=PART(F,1);

        G := A

We can use PART to pick out subparts in one step. The last two commands can be combined as

    G:=PART(EE,2,1);

        G := A

It's time for a more realistic example. Suppose we see a computation result

$$EE := (7*(15*A*B^N + 37*C^2*D^5 + P*Q))$$
$$/(13*(A*B^2 + P^2*Q))$$

and we want U to be the part of the numerator from "15" to "Q⁵". The entire numerator (including the factor 7 in front) is Part 1. Subpart 1 of that is the "7"; subpart 2 is what we want. We could type it in; we could use NUM and divide the result by 7; or we can use PART:

`U:=PART(EE,1,2);`

$$U := 15*A*B^N + 37*C^7*D^5 + P*Q$$

Or suppose we just want the term beginning with the 37, without typing it in by hand:

`V:=PART(EE,1,2,2);`

$$V := 37*C^N*D^7$$

As a special convenience when selecting a part of a long expression, one has the option of counting parts from the end of the expression instead of the beginning. The last Part can be referred to as Part -1; the next-to-last Part, as Part -2; and so on.

To understand PART a little better it is necessary to consider the syntax of REDUCE expressions. Each expression (other than a variable or constant by itself) has a Main Operator (sometimes called Main Connective). For example, in EE := A*B*C*D + P*Q + W the expression is the sum of three terms. The Main Operator is PLUS. PLUS can relate two or more terms. Each term is a Part: here, we have Parts 1,2,3.

In Part 1, A*B*C*D, the Main Operator is TIMES. TIMES

PART and setting a PART   119

can relate two or more factors. Each factor, here, is a subpart of EE.

If the expression has a denominator, the Main Operator is QUOTIENT. In contrast to PLUS and TIMES, QUOTIENT must relate exactly two Parts: the Numerator and the Denominator. Thus if EE is a fraction, PART(EE,1) and PART(EE,2) are exactly equivalent to NUM(EE) and DEN(EE).

Minus signs require special treatment. The reader may guess that the Main Operator in A - B is MINUS. If he does so he is mistaken! PART does not recognize subtraction as an operator. To PART, the expression A - B looks like A + (-B). The Main Operator is PLUS. Part 2 is -B, an expression whose Main Operator is a one-place operator MINUS, and whose Part 1 is the "B".

The reader doesn't have to accept these descriptions of the Main Operator on faith. Part 0 of any expression is the Main Operator. For example,

    EE:=A-B;

            EE := A - B

    PART(EE,O);

            PLUS

    PART(EE,2);

            - B

    PART(EE,2,O);

            MINUS

If a symbol has been declared OPERATOR, any expression beginning with it has the operator as Part 0, and as many Parts as there are operator places shown:

```
OPERATOR YY,ZZ;

EE:=YY(A) + ZZ(B,C);

       EE := YY(A) + ZZ(B,C)

PART(EE,1,0);

       YY

PART(EE,2,0);

       ZZ

PART(EE,2,1);

       B

PART(EE,2,2);

       C
```

2.11.2. Setting a PART

PART enables us to examine any part of an expression. A curious notation using PART and ":=" enables us to change the part. We illustrate this.

```
P:=4*B**5 + 2*A**3 + 7*C**8;

              3       5       8
       P := 2*A   + 4*B   + 7*C
```

Now suppose we want Q to be just like P but with the exponent 5 changed to 10. The term 4*B**5 is part 2 of P (as printed). Part 1 of that is the 4; part 2 the B**5. The 5 we want to change is part 2 of that. Let us use PART to verify this analysis:

```
PART(P,2,2,2);

       5
```

Correct! Now we construct Q:

PART and setting a PART

    Q:=(PART(P,2,2,2) := 10);

$$Q := 2*A^3 + 4*B^{10} + 7*C^8$$

Q is like P except that it has a 10 where P had the 5. P itself is not changed:

    P;

$$2*A^3 + 4*B^5 + 7*C^8$$

How to use PART to change something should be clear from this example. We place in parentheses the PART symbolism for the part to be changed; next, write ":=" which, in this context, is <u>not</u> an assignment operator; and last, the expression to be inserted in place of that part.

Note that "changed" must not be taken literally. Nothing is changed by PART(...):=...; a new expression is formed that is like the original but for a single difference. If we DO want to change the original expression, P, we have to use an assignment (a second, "real", :=);

    P:=(PART(P,2,2,2) := 10);

The printing order of the terms of the modified expression, or indeed its entire structure, could be changed by using PART in this way. For example:

    Q:=(PART(P,1,2,1) := W);

$$Q := 4*B^5 + 7*C^8 + 2*W^3$$

The term that was 2*A**3 has migrated to the end. When carrying out a succession of part-changing operations, always print out the expression after each change in order to locate correctly the part to be changed. Alternatively, use PART before each each change to verify that the right part is about to be changed (as we did in the first example).

With care PART can even be used to change Part 0, the Main Operator of an expression.

```
P:=A+B+C+D;
        P  :=  A  +  B  +  C  +  D
Q:=(PART(P,0)  :=  *);
        Q  :=  A*B*C*D
```

Notice that a star * (meaning "multiply these four expressions", was used to replace, simultaneously, the three +'s (meaning "add these four expressions"). This change worked because internally P is the "PLUS" of A, B, C, D, and Q is the "TIMES" of the same four variables.

The Main Operator can even be changed to a user-introduced operator:

OPERATOR H$

```
G:=(PART(P,0)  :=  H);
        G  :=  H(A,B,C,D)
```

We conclude this section by pointing out that the expression (PART(...):=...) is a legitimate expression which can be used in other ways than just to be assigned directly to a variable. For example, we could write

```
R:=23*(PART(P,2,2,2)  :=  10)**2/(PART(P,2,2,2)  :=  56);
```

We'll even confess that the parentheses around the (PART(...):=...) weren't necessary in the simpler examples we gave at the beginning of this subsection, but were used to emphasize that (PART(...):=...) is really an expression in spite of its appearance.

3. Setting Modes and Options

The user has some control over how REDUCE operates internally, and also over the form in which results are printed out. Most of these options are set or changed by using the ON and OFF commands. Instead of Options we sometimes speak of Modes, or speak of the ON and OFF commands as setting Switches. Some options (modes, switches) are normally ON; the user changes them by issuing an OFF command. The option then remains OFF until the user returns it to normal by issuing an ON command. For other options the normal mode is OFF, so the sequence of commands would be reversed.

In this chapter we will describe many of the available options, and of course indicate for each whether the normal mode is ON or OFF. There are relatively few exercises. The reader should familiarize himself with the available options, so that he will be aware of the possibilities when he is in the midst of a calculation he is carrying out with the help of REDUCE.

We caution the reader that there is no check in REDUCE that the name of the switch being turned ON or OFF is correctly typed. A mistyped ON or OFF command -- such as **OFF EZP$** instead of **OFF EXP$** -- is simply ignored. (In one version of REDUCE, one may get a ... **declared FLUID** message the first time he attempts to set a nonexistent switch.)

3.1. EXP

REDUCE normally expands -- multiplies out -- any expression which it is given.

     A := (X+Y)**5;

$$A := X^5 + 5*X^4*Y + 10*X^3*Y^2 + 10*X^2*Y^3 + 5*X*Y^4 + Y^5$$

Not only are six terms printed out, but the internal representation of the value of A contains these six terms.

If we had the 100th power instead of the 5th power, there would be 101 terms. If we set $A:=(X+Y)**100$ and $B:=(X+Y)**101$, and asked REDUCE to calculate $A*(X+Y)-B$, REDUCE would multiply the 101 term expression A by $(X+Y)$, and from the 102 term result subtract, term by term, the 102 terms of B, and of course finally get the answer zero. All this takes much time and much -- possibly too much -- computer memory.

If we actually printed out the value of A, we would have to be very observant to notice that A is exactly the 100th power of $(X+Y)$.

The picture changes completely if we change the switch EXP from its normal ON mode to OFF.

OFF EXP$

$A := (X+Y)**5;$

$A := (X + Y)^5$

This time the power is not multiplied out, and the value of A is stored more compactly.

Let's continue:

ON EXP$

$B := A\$$

Since the EXP mode is now ON, B receives as value the expanded form of the value of A. This will be evident if we print the value of B even with EXP set back to OFF:

EXP

```
OFF EXP;

B;
```

$$B := X^5 + 5*X^4*Y + 10*X^3*Y^2 + 10*X^2*Y^3 + 5*X*Y^4 + Y^5$$

Now we are ready to appreciate the reason the normal mode of EXP has been chosen to be ON. Does REDUCE know that A and B are equal? Remember that the mode of EXP is now OFF.

```
A - B;

    0
```

Apparently it does. But how? To see that A, in the form $(X+Y)**5$, and B, expanded out, are actually equal it has to (essentially) expand out the power anyway, even though the EXP mode is OFF, so that it can verify that all terms cancel. REDUCE tries to be careful that any expression that is really zero is recognized as such.

Continuing with the example we get an even more impressive illustration of the difficulties under which REDUCE labors in the OFF EXP mode. Let us change B by removing one of its terms:

```
ON EXP;

B := B - 5*X*Y**4;
```

$$B := X^5 + 5*X^4*Y + 10*X^3*Y^2 + 10*X^2*Y^3 + Y^5$$

Does REDUCE, with EXP off, know that A and B are almost equal?

```
OFF EXP;

A - B;

    5*X*Y^4
```

Even in the OFF EXP mode, if the value of an expression (like our A-B) would be much simpler expanded than with its parts unexpanded, REDUCE may produce an expanded or partially expanded form. Thus we can only say that in OFF EXP mode expressions are "usually" left unexpanded.

We now give another example. In the normal ON EXP mode, we assign a value to W:

W := (A + B)*(X + 5);

W := X*A + X*B + 5*A + 5*B

As expected, the value of W is expanded out. Now we change the mode:

OFF EXP$

W;

(A + B)*X + 5*A + 5*B

When not forced to expand W, REDUCE found it more efficient to factor out X from two terms of the expression. One would have to examine the inner workings of REDUCE quite intensively to determine why REDUCE didn't also factor out 5 from the other two terms. In fact, the same problem run on another version of REDUCE produced the answer (A + B)*(X + 5). This shows that if EXP is OFF, the form of the output can not be readily predicted!

3.2. GCD

When given a fraction to simplify, REDUCE normally exerts only a limited effort at cancelling common factors from the numerator and denominator. Identifying all common factors is quite time consuming.

Observe the following computation:

GCD

```
AA := XX*RR$

BB := YY*RR$

AA/BB;
```

$$XX/YY \qquad \text{\% common factor RR cancelled}$$

```
RR := SS+1$

AA/BB;
```

$$(XX*(SS + 1))/(YY*(SS + 1))$$
$$\text{\% common factor SS+1 not cancelled}$$

The common factor RR was "easy" to find, and was cancelled. The common factor (SS+1) was "hard" to find, and was not cancelled.

By issuing the command ON GCD$ the user can instruct REDUCE to put forth the effort required to find and cancel all common factors (the Greatest Common Divisor):

```
ON GCD$

AA/BB;
```

$$XX/YY$$

Naturally, the command OFF GCD$ should be typed in as soon as the slow ON GCD mode is no longer wanted.

The reader who tries to duplicate the example in which the common factor (SS+1) didn't cancel may or may not succeed. Whether (SS+1) is an easy factor for REDUCE to find, or not, and so whether it is cancelled in OFF GCD mode, depends on the internal representation of AA and BB. (And this depends in part on the order in which variables were introduced during the REDUCE run.) The KORDER command, which was mentioned in an earlier chapter and which will be explained in more detail later in this chapter, starting on page 149, can control the aspect of the internal representation that is relevant in this example. To make sure this example works as shown, KORDER SS$ should be inserted before it. And if one wants to make sure the example doesn't work as shown, he should insert **KORDER XX\$** before it: then the (SS+1) factor will be easy for REDUCE to find, and the **ON GCD\$** would be unnecessary.

We may sometimes make use of the fact that OFF GCD is available (and even is the normal mode). If we need to work with a variable X defined to be

X := (F**10 - G**10)/(F-G);

$$X := F^9 + F^8*G + F^7*G^2 + F^6*G^3 + F^5*G^4 + F^4*G^5 + F^3*G^6 + F^2*G^7 + F*G^8 + G^9$$

we see that the expanded form is much harder to read and comprehend than the given fractional form. Since the denominator F-G does actually divide the numerator it makes no difference to the output we get whether the mode GCD is ON or OFF. But we can change the denominator by introducing another factor, a "new" variable, say ABC. Its presence allows the fractional form to remain, in the OFF GCD mode:

OFF GCD\$ % (in case we aren't sure)

X := (F**10 - G**10)/(ABC*(F-G));

$$X := (F^{10} - G^{10})/(ABC*(F - G))$$

When we really need X we can set **ABC:=1** to remove the ABC.

(If you try this example and find X expanded out, try first entering **KORDER F\$**.)

The fact that common factors aren't always cancelled doesn't usually affect REDUCE's ability to recognize zero expressions. After all, the numerator of a zero fraction is zero whether or not extra factors are present. We can see this by continuing with the A/B example:

GCD

```
OFF GCD$                    % (to make sure)

AA/BB;
```

$$(XX*(SS + 1))/(YY*(SS + 1))$$

```
XX/YY - AA/BB;
```

0

But introduce an operator symbol, preventing REDUCE from forming a single fraction with zero numerator:

OPERATOR H$

```
Z := H(XX/YY) - H(AA/BB);
```

$$Z := - H((SS*XX + XX)/(SS*YY + YY)) + H(XX/YY)$$

Not shown as zero! ON GCD will come to the rescue:

ON GCD$

```
Z;
```

0

The combination ON GCD and OFF EXP provides a limited ability to factor polynomials, which while limited is much faster than the full factoring facility provided by FACTORIZE (page 68) or the equivalent ON FACTOR (page 138). Any polynomial can theoretically be factored into non-repeated factors, twice-repeated factors, three-fold factors, and so on. In the ON GCD/OFF EXP combination of modes polynomials will be printed as the product of the non-repeated factors, times the square of the product of the twice-repeated factors, times the cube of the product of the three-fold factors, etc. Some of these products may be further factored.

```
A := (A1+A2)*(A1+B2)*(C1+C2)**2*(D1+D2)**2
     *(E1+E2)**3*(E1+F2)**3$
     % Note the "$", not ";" -- you don't want
     % to watch this printing out expanded!
     % It has 576 terms!

OFF EXP$
ON GCD$

A;
```

$$((A1^2 + A2*B2) + (A2 + B2)*A1)$$
$$*(C1 + C2)^2 *(D1 + D2)^2$$
$$*(E1^2 + E1*E2 + E1*F2 + E2*F2)^3$$

Don't expect this calculation to be very rapid: when tested on a DEC 20 mainframe computer it took about 15 seconds of computer time, while a Zenith-158 microcomputer required nearly 10 minutes. (With the last cubic factor omitted from A, the Zenith used 80 seconds.) The answer has a first factor which is $(A1+A2)*(A1+B2)$, partly multiplied out and with some collecting of terms. The second factor, which represents $((C1+C2)*(D1+D2))^2$, is completely factored. The third factor, representing the two cubic factors in the given expression, is printed as the cube of the multiplied-out product. The reader should not expect to be able to predict exactly how much factoring he will get.

The user will have to learn by experience when ON GCD is desirable. For example, some matrix computations with matrices with symbolic entries have much simpler-looking answers if the mode setting combination ON GCD and OFF EXP is used.

3.3. LCM

When two fractions are added they are placed over a common denominator. REDUCE normally forms the Least Common Multiple of the original denominators. To do this it first finds their Greatest

Common Divisor, even if -- as is usual -- the GCD mode switch is OFF.

To suppress even this limited use of the GCD algorithm the LCM switch, which is normally ON, can be turned OFF. To illustrate:

    A:=(X + 1)*(X + 2)$

    B:=(X + 1)*(X + 3)$

    C := 1/A + 1/B;

$$C := (2*X + 5)/(X^3 + 6*X^2 + 11*X + 6)$$

The common factor (X+1) was eliminated. Compare this with

    OFF LCM;

    C := 1/A + 1/B;

$$C := (2*X^2 + 7*X + 5)/(X^4 + 7*X^3 + 17*X^2 + 17*X + 6)$$

Exercise 3.3.1.

If C is formed under the OFF LCM condition as shown immediately above, should

    ON LCM$

    C;

be expected to show C simplified? If not, what can be done to simplify C?

3.4. MCD

REDUCE normally treats denominators as (no surprise here) denominators:

X := 1/(A*U);

$$X := 1/(A*U)$$

If an expression is a sum of two fractions REDUCE will put the sum over a common denominator (normally the least common denominator):

Y := X + 1/(B*U);

$$Y := (A + B)/(A*B*U)$$

Z := X + 1/(B*V);

$$Z := (A*U + B*V)/(A*B*U*V)$$

If the switch MCD, which is normally ON, is set to OFF then denominators are treated as negative powers and very little combining takes place.

OFF MCD$

X := 1/(A*U);

$$X := A^{(-1)} * U^{(-1)}$$

Y := X + 1/(B*U);

$$Y := U^{(-1)} * (A^{(-1)} + B^{(-1)})$$

Z := X + 1/(B*V);

$$Z := A^{(-1)} * U^{(-1)} + B^{(-1)} * V^{(-1)}$$

The OFF MCD mode should be used with caution. In this mode expressions that are equal to zero are not always recognized as such:

MCD

```
R := S+1$

A := R/R - 1;
                (-1)              (-1)
        A := (S + 1)   *S + (S + 1)    - 1
```

The fact that A is zero is not recognized while in OFF MCD mode.

3.5. RESUBS

At the very beginning of our study of REDUCE we learned that the value of a variable is traced down until it's expressed entirely in terms of constants and clear variables. It is possible to ask REDUCE to stop after a single stage of evaluation, by setting to OFF the switch RESUBS.

Consider the following:

A:=B$

B:=C$

C:=D$

A;

      D

This is what we would expect. But observe:

OFF RESUBS;

A;

      B

B;

      C

C;

      D

In the OFF RESUBS mode, only a single stage of evaluation takes place each time.

Before too much dependence is put on this, we must warn the reader that a thorough knowledge of the inner workings of REDUCE is needed to determine reliably what a "single evaluation stage" consists of. But generally, printing a variable A in the OFF RESUBS mode shows its "assigned value", which was defined in Chapter 1 to be what the value of the right-hand side of the := was at the time the most recent assignment to A was made.

3.6. ORDER

By this time the reader must have already noticed that printing X+Y results in the output X+Y, and that printing Y+X also results in the output X+Y. REDUCE prints expressions in a standard form, which includes a standard ordering of the variables used in the expression. In some implementations of REDUCE, any expression containing only single-letter variables is printed in an order deriving from the order of the letters in the alphabet. Most other variables are printed in an order partly deriving from the order in which they were first introduced during the current REDUCE session.

Exercise 3.6.1.

Start a new REDUCE session by printing out the value of CAT + DOG; then print out the values of CAT * DOG and of DOG * CAT. Start another REDUCE session by printing out the value of DOG + CAT; then print out the values of CAT * DOG and of DOG * CAT. Does the last assertion made in the text above apply to your version of REDUCE?

If the user wants to change the order in which REDUCE prints out variables, he can use the ORDER command.

```
ORDER Y,X,B,C$
```

tells REDUCE that in printing expressions containing these variables the order should be based on the indicated ordering of the variables.

Variables not listed will come after any that are listed.

  B + C + X + Y + X*B + (A+B+C)*Y + A + CAT;

  Y*B + Y*C + Y*A + Y + X*B + X + B + C + A + CAT

If two ORDER commands have been entered, the result is the same as if the variables had been listed in a single sequence.

ORDER Y,X,B,C$

followed immediately or later by

ORDER P,Q,R$

orders the variables as Y,X,B,C,P,Q,R from then on.

If a later ORDER repeats a variable listed in an ORDER command earlier, the second occurrence is what counts:

ORDER Y,X,B,C$

followed immediately or later by

ORDER P,X,Y,Q,R$

orders the variables as B,C,P,X,Y,Q,R.

The special command **ORDER NIL$** causes the effect of all prior ORDER commands to be cancelled.

Exercise 3.6.2.

Define X to be $A1*A2^3 + B1^4*B2$, and print it out in three different orders.

Exercise 3.6.3.

Suppose **ORDER B,C,D,E,A$** has already been specified, and the user now decides he wants A first. What can he do? Solve this both without and with using **ORDER NIL$**.

ORDER can also be used to specify the ordering of specific

operator forms such as H(A).

```
OPERATOR H$
ORDER D,C,B,A$

W := A + B + C + D + H(A) + H(B);

    W := D + C + B + A + H(A) + H(B)
```

(Notice that H(A) happened to come before H(B) even though B comes before A.) Now let us move H A and H B to the front (the parentheses can be omitted as usual):

```
ORDER H A,H B,D,C,B,A$

W;
```

$$H(A) + H(B) + D + C + B + A$$

There is no way to specify that all H( ) forms should be in some specific point in the ordering. Each H( ) of interest must be listed individually in an ORDER command, if its position is to be controlled.

The KORDER command has already been mentioned several times, and will be introduced formally on page 149. It, too, has something to do with ordering, but this ordering is the one to be used for the internal representation of expressions. The ordering for output is different: it completely ignores any KORDER that may have been specified (except in OFF PRI mode), and depends only on ORDER.

3.7. FACTOR command

The somewhat misnamed FACTOR command gives us a different kind of control over the form of output REDUCE gives us.

FACTOR command

$$W := X**2*Y**7*A + 5*X**2*Y**3 + 7*X*C - 9*Y$$
$$+ X**2*Y**3*B - X + Y*D + 123 + P;$$

$$W := D*Y + A*X^2*Y^7 + 7*C*X + B*X^2*Y^3 + P + 5*X^2*Y^3 - X - 9*Y + 123$$

FACTOR X,Y$

W;

$$X^2*Y^7*A + X^2*Y^3*(B + 5) + X*(7*C - 1) + Y*(D - 9) + P + 123$$

All terms with the same combination of powers of X and Y are grouped together in printing. Last are the terms containing neither X nor Y.

If we had written FACTOR Y,X$ the Y's would have been printed before the X's in each term, and the order of the terms would have been different.

The effect of another FACTOR command in the course of the REDUCE session is to add more variables to the collection of variables to be "factored out". This frequently reduces the amount of grouping, because it increases the number of variables whose powers must match in each group.

The command REMFAC is used to remove variables from the "factoring" list. REMFAC X,Y$ would cancel the effect of the FACTOR X,Y$ completely. REMFAC X$ leaves Y on the "factoring" list:

REMFAC X$

W;

$$Y^7*A*X^2 + Y^3*X^2*(B + 5) + Y*(D - 9) + 7*C*X + P - X + 123$$

Operators can also be listed in FACTOR commands. If SIN is listed, then SIN(...) and its powers, with any expression at all in the parentheses, are "factored out". If an operator is listed with a specific expression in the following parentheses, only that specific form and its powers are "factored out".

OPERATOR H$

```
W :=   SIN(U) * X1 + SIN(V) * X2
     + SIN(U) * X3 + SIN(V) * X4
     + (H(A)+1)**2 * (X5+X6)
     + (H(B)+1)**2 * (X7+X8)$
```

FACTOR SIN, H(A)$

W;

$$SIN(U)*(X1 + X3) + SIN(V)*(X2 + X4) + H(A)^2*(X5 + X6)$$
$$+ 2*H(A)*(X5 + X6) + H(B)^2*X7 + H(B)^2*X8 + 2*H(B)*X7$$
$$+ 2*H(B)*X8 + X5 + X6 + X7 + X8$$

One can see that SIN(U), SIN(V), H(A), and $H(A)^2$ were "factored out", but H(B) and $H(B)^2$ weren't.

3.8. FACTOR switch

We have already looked at the FACTOR <u>command</u>, which is used in commands like FACTOR X,Y$ and causes grouping of terms in output in a certain way. The FACTOR <u>switch</u>, set by writing ON FACTOR$ and restored to normal by OFF FACTOR$, is something quite

FACTOR switch

different.

The command ON FACTOR$ causes REDUCE to attempt to factor every expression it encounters, using the FACTORIZE function whose use was described on page 68. It turns the EXP mode OFF, and prints answers in as nearly factored a form as it is able to. Since factoring is difficult, this mode of operation is of course generally much slower than the normal mode. The exception is if the factored expressions formed early in a computation are so much shorter than they would be otherwise that later stages of the computation run faster because they are operating on simpler data.

In some implementations of REDUCE, the command LOAD "FACTOR"$ must be given before the first use of ON FACTOR in a session, unless of course this was already done in order to use FACTORIZE.

Let's start our demonstration with two expressions input in the normal (ON EXP) mode:

A := (X+Y)*(P+Q);

A := X*P + X*Q + Y*P + Y*Q

OPERATOR H$

AA := H(X**2 - Y**2);

$$AA := H(X^2 - Y^2)$$

Now turn on the FACTOR mode:

ON FACTOR$

A;

(X + Y)*(P + Q)

AA;

$$H(X^2 - Y^2)$$

The first expression factors. We discover that FACTOR doesn't act inside operator symbols like our H( ).

```
B := X**2 - 25*Y**2;
```

$$B := (X + 5*Y)*(X - 5*Y)$$

```
C := X**2 - 26*Y**2;
```

$$C := X^2 - 26*Y^2$$

We see that FACTOR doesn't factor if it would have to introduce SQRT(26) or the like.

```
D := X**2 + 1;
```

$$D := X^2 + 1$$

We see that it doesn't introduce I, that is, SQRT(-1), either.

```
EE := B**2 + 9;
```

$$EE := (X + 5*Y)^2 *(X - 5*Y)^2 + 9$$

This is unexpected. EE doesn't factor, but a part of it does. Since EXP is OFF when we are in FACTOR mode, this partly factored result is printed.

Two more examples to round out the list:

FACTOR switch

```
F := B + 9*Y**2;

        F := (X + 4*Y)*(X - 4*Y)

X**15 - 1;
```

$$(X^8 - X^7 + X^5 - X^4 + X^3 - X + 1)$$
$$*(X^4 + X^3 + X^2 + X + 1)$$
$$*(X^2 + X + 1)*(X - 1)$$

The command **OFF FACTOR$** terminates the factoring mode, and also turns the EXP mode on (even if it hadn't been on when the **ON FACTOR** command was issued).

If the reader is curious about how FACTOR works, he can issue the command **ON TRFAC$** which will cause considerable explanatory printout to be produced during factoring. Of course **OFF TRFAC$** will silence FACTOR.

3.9. DIV

Suppose an expression with a denominator is to be printed, and the denominator contains simple factors such as numbers like 12 and/or (clear) variables like A. If the DIV switch is changed to ON, instead of these simple factors appearing in the denominator in the printout they appear in the numerator as fractions like 1/12 or as variables to negative powers like $A^{(-1)}$.

```
W := (P + Q)$

Y := 12*A*B$

ON DIV$

W/Y;
```

$$1/12 \ast P \ast A^{(-1)} \ast B^{(-1)} + 1/12 \ast Q \ast A^{(-1)} \ast B^{(-1)}$$

```
B := U + V$

W/Y;
```

$$(1/12 \ast P \ast A^{(-1)} + 1/12 \ast Q \ast A^{(-1)})/(U + V)$$

(U+V) not being a simple factor, it stays in the denominator.

Operator expressions like H(A) or SQRT(U+V) are considered simple, and so they, too, are brought into the numerator:

```
B := SQRT(U + V)$

W/Y;
```

$$1/12 \ast SQRT(U + V)^{(-1)} \ast P \ast A^{(-1)}$$
$$+ 1/12 \ast SQRT(U + V)^{(-1)} \ast Q \ast A^{(-1)}$$

Do not confuse ON DIV with OFF MCD. The former affects the internal representation of fractions (and only secondarily influences the output form). With OFF MCD nothing is left in the denominator, not even sums like (U+V). As was pointed out in the section on MCD, in that mode REDUCE overlooks the fact that $(U+V)^{-1} \ast (U+V)$ equals 1, and so answers may be overly complicated at times. This doesn't arise with ON DIV, which only affects the appearance of the output and not the internal representation of expressions.

3.10. RAT

The RAT mode, which is normally OFF, is relevant only when a FACTOR command is in effect <u>and</u> a denominator is present.

In the ON RAT mode the expression is printed with the denominator repeated with each of the terms into which the FACTOR operation divides the expression. If any of these terms have an easy-to-find factor in common with a factor in the denominator (in the sense discussed in the section on the GCD switch), the common factor is cancelled in printing. (If the GCD mode is also ON, all common factors, not just easy ones, are cancelled in printing.)

Note that ON RAT affects only the way expressions are printed. It doesn't change the way they are stored internally.

Exercise 3.10.1.

Demonstrate that both the **FACTOR A$** and the **ON RAT$** are necessary in order to get the result shown below. (Why does the **FACTOR A$** seem to do nothing?)

```
FACTOR A$
ON RAT$

W := (A*B + C*D)/B;

     A + (C*D)/B
```

3.11. ALLFAC

In printing expressions REDUCE normally factors out any constants or variables that divide every term.

```
W := 5*A**3*Q + 20*A*R**2 + 35*A;
```

prints out as

$$W := 5*A*(4*R^2 + A^2*Q + 7)$$

because every term is divisible by 5 and by A. But if we add 1 nothing factors out (because the term 1 is divisible by neither 5 nor by A):

```
W + 1;
```

$$20*R^2*A + 5*A^3*Q + 35*A + 1$$

Only constants, variables, and operator forms such as SIN(A) or SQRT(B+C) are factored out. Sums and differences are not. For example, in

```
W := (A + B)*(X + 5);
```

the (A+B) is not factored out in the printout:

$$W := A*X + 5*A + B*X + 5*B$$

Exercise 3.11.1.

Verify that the operator form SQRT(B+C) factors out if it appears in every term of an expression.

If one wants to examine the coefficient of a term, this automatic factoring is a nuisance, since one has to combine the coefficient that appears in the term itself with the coefficient, if any, in front of the whole expression. Setting the mode ALLFAC to OFF supresses this feature.

```
OFF ALLFAC$

W;
```

$$20*R^2*A + 5*A^3*Q + 35*A$$

Of course ON ALLFAC$ restores this option to normal.

We note that the factoring that is carried out in ON ALLFAC mode is done separately for the numerator and the denominator of a fraction.

ALLFAC

If use of the FACTOR command, say **FACTOR A$**, has introduced parentheses into the printed form of an expression (for example, to collect all the terms containing A**2) then the ALLFAC factoring is done separately in each of these parentheses.

If **FACTOR A$** and **ON ALLFAC$** are both in effect, and all terms contain at least the first power of A, the FACTOR A output form takes precedence: that is, no A is factored out of the entire expression although mathematically it could be.

Exercise 3.11.2.

(Review of ORDER, FACTOR, REMFAC, DIV, RAT, and ALLFAC.) The following are printouts of the value of one and the same variable, under the influence of one or more of the above commands. Discover what commands were used. (Hint: in each case exactly one or two commands were used to get from each output form to the next.)

$$(X*(2*A*X*Y^2 + 4*A*X*Y + Y^2 + Z))/(2*A)$$

$$(2*X^2*A*Y*(Y + 2) + X*(Y^2 + Z))/(2*A)$$

$$(2*X^2*Y^2*A + 4*X^2*Y*A + X*Y^2 + X*Z)/(2*A)$$

$$X^2*Y*(Y + 2) + A^{(-1)}*X*(Y^2 + Z)/2$$

$$X^2*(Y^2 + 2*Y) + A^{(-1)}*X*(Y^2 + Z)/2$$

$$X^2*(Y^2 + 2*Y) + A^{(-1)}*X*(1/2*Y^2 + 1/2*Z)$$

3.12. LIST

The command **ON LIST$** causes each term in an expression to print out on a separate line, so the terms are "listed" in a column. Each plus sign and minus sign begins a new line.

```
ON LIST$
W := A*B*C + D*E*F - G*H*I;
    W := A*B*C
        + D*E*F
        - G*H*I
```

This output form is not quite as satisfactory if fractions are present:

```
(P+Q+R)/(A-B+C);
        (P
          + Q
          + R)/(A
                 - B
                 + C)
```

or if ALLFAC or FACTOR introduces parentheses:

```
5*A*(P+Q+R);
        5*A*(P
              + Q
              + R)
```

Plus and minus signs within operator symbols do not cause new lines:

```
OPERATOR H$
H(P+Q+R) + H(A-B+C);
        H(R + P + Q)
          + H(A - B + C)
```

3.13. NERO

In the ON NERO mode, assignments of the value 0 to a variable do not print. (NERO = "No Zero".) This applies both to printouts resulting automatically from use of the semicolon as delimiter, and to printouts resulting from WRITE.

```
ON NERO$

U := 1;
        U := 1
U := 0;
W := U*(B+C);
<<WRITE W:=2; WRITE W:=0; WRITE W:=5;>>$
        W := 2
        W := 5
```

Notice that there is no output resulting from the second and third assignments, even though they end with a semicolon, because in both of those assignments the value being assigned to the variable is zero. Nor is there any from the middle one of the three WRITE's, for the same reason.

ON NERO suppresses only <u>assignment</u> printouts. If no ":=" occurs the printout is normal:

```
U*(B+C);

        0
```

The real use of ON NERO is not with single assignments, but with assignments within a repeating structure. In the example which follows, A is an array which is mostly zero. It is being copied to the B array. The printout records only the non-zero array items copied.

```
FOR I:=0:10 DO
    WRITE B I := A I$

    B(3) := ABC

    B(10) := DEF + 123
```

There is an exception to the rule that ON NERO applies only to assignments. ON NERO also suppresses printing of the zero elements when printing out a matrix even if the user didn't specify an assignment. (This is because while REDUCE prints a matrix it assigns it to a ficticious matrix MAT.) We illustrate this by printing out the non-zero elements of a matrix M previously constructed:

```
M;

    MAT(1,1) := 1

    MAT(1,3) := 5

    MAT(2,2) := 2

    MAT(2,3) := 7
```

3.14. NAT, FORT

The "natural" mode of REDUCE output places exponents on the line above the main line of the expression. The command OFF NAT$ changes output formatting to be similar to input: Exponents are printed in line with the rest of the expression, preceded by the symbol "**".

```
OFF NAT$

W := (5*A**3*B**6 + P)/Q**2;

    W := (5*A**3*B**6 + P)/Q**2$
```

This output form is primarily intended for output to files to be read into REDUCE later. See the OUT command explained in the Running REDUCE chapter, on page 306.

OFF NAT is intended to produce output in a form suitable for later input into REDUCE. If output is desired for insertion into a FORTRAN program, the command ON FORT$ should be used instead. This is discussed more fully on page 308.

3.15. PRI

The formatting, factoring, re-ordering, and so on which have been described in several of the preceding sections take considerable time. If the REDUCE user doesn't care about any of these features, and wants to speed up printing, he can use the command OFF PRI$ to instruct the output package to display expressions in a form closely reflecting the form in which they are stored internally. The output is still in familiar algebraic form, but the ordering and grouping of terms may seem eratic.

OFF PRI is not only useful for saving time. It is helpful if one wants to examine the internal organization of expressions without going to the extreme of descending to the RLISP level. We shall use OFF PRI in the next section, for example, to help explain the KORDER command. The reader will find examples of OFF PRI output there. Another example is given in the Case Studies chapter, on page 279, where it is used in tracking down a "bug" in REDUCE.

3.16. KORDER

In the section on ORDER we have already explained that the variables used during a REDUCE session are internally arranged in an order that is in part fixed for once and for all, and in part a consequence of the order in which they were introduced during the session. We shall call this the default ordering.

If he needs to know, the REDUCE user can determine how two variables compare under the default ordering by using the Boolean function ORDP. ORDP(X,Y) is true if X is earlier than Y in the default ordering, and false if not. Since ORDP is a Boolean function, we can't print out its value directly, but must use it in some IF ... THEN (or similar) environment, for example as in

```
PROCEDURE ORDFP(X,Y);
IF ORDP(X,Y) THEN X ELSE Y;
```

Evaluating ORDFP(X,Y) yields whichever of the two arguments, X or Y, is earlier in the default ordering. The two arguments must be clear. (The topic of Procedures will be formally presented in Chapter 4.)

Both the internal representation of expressions, and the order in which the variables print during output, initially depend on the default ordering. We have already seen how this ordering can be changed, for printing purposes, by the ORDER command. For controlling the internal representation it can be changed by using the KORDER command. The two are independent: ORDER has no effect on the internal representation, and KORDER has only minor effect on output form (except in the OFF PRI mode). Neither affects which variable will be the one actually replaced when a LET rule like **LET AA+BB=1000** is processed -- that decision is controlled exclusively by the default ordering. To summarize, there are three orderings of variables that may be of concern: the default ordering, the ordering set up by the ORDER command, and the ordering set up by the KORDER command. At the end of this section we will tabulate which ordering is relevant to which aspects of REDUCE's behavior.

A command like **KORDER X,Y,W$** tells REDUCE to consider X to be the "first" of all variables, Y next, W next, and all other variables following in their default order.

Recall that repeated ORDER commands have a cumulative effect. The situation for KORDER is different: any KORDER command cancels all previous KORDER commands. No command analogous to REMFAC is necessary. To return to the default ordering, the ordering that existed before any KORDER commands, just issue the special command **KORDER NIL$**.

A polynomial in several variables is stored as a polynomial in the "first" of its variables, ("first" in KORDER ordering), with coefficients that are polynomials in the others. We can see this by using OFF PRI.

OFF PRI$

EE := X**2*S**2 + X**3*S**2 + X*S + X + 123$

KORDER S$

EE;

$$(X^3 + X^2)*S^2 + X*S + X + 123$$

This output (which is the same as we would have gotten without the KORDER S, actually) can be seen to be arranged in order of powers of S: first the S**2 terms; then the S term; finally the two terms not containing S at all.

Let us change the KORDER:

KORDER X$

EE;

$$S^2*X^3 + S^2*X^2 + (S + 1)*X + 123$$

Now the output is arranged in order of decreasing powers of X.

Exercise 3.16.1.

Study and explain:

OFF PRI$

KORDER X,B,Y,A$

C:=(X + Y)*(A + B);

$$(B + A)*X + Y*B + A*Y$$

All this would be interesting but of no significance if the internal structure of expressions never mattered. But we have already pointed out two situations in which it does.

The first was LET sum = ..., but in that context KORDER is

ineffective. What variable will be selected out depends only on the default ordering.

The second of the prior references to KORDER concerned the common factors that are cancelled from fractions in the normal OFF GCD mode. We reproduce an example from page 126:

```
AA := XX*RR$
BB := YY*RR$
RR := SS+1$
AA/BB;
```

$$(XX*(SS + 1))/(YY*(SS + 1))$$

Let us use OFF PRI to see why the (SS+1)'s didn't cancel.

OFF PRI$

AA;

$$XX*SS + XX$$

We see that AA is arranged in powers of SS (as the first power followed by the zero'th). The factor SS+1 is obscured. Let us use KORDER to remedy this:

KORDER XX$

AA;

$$(SS + 1)*XX$$

Much better! Now AA is arranged in powers of XX (there is only one, the first), and the SS+1 factor is obvious.

AA/BB;

$$(XX) / (YY)$$

REDUCE could detect the common factor SS+1 even without using the general GCD algorithm.

Exercise 3.16.2.

First enter the assignment S := (A-B)/(C-D); to get the result S := (A - B)/(C - D). Then enter the appropriate combination of KORDER and ORDER commands so the value of S prints out in the form (B - A)/(D - C).

As promised, we tabulate the situations in which each of the three orderings discussed are relevant.

- default ordering: basis for ORDER and KORDER; selection of variable picked out in LET rules such as LET A+B = ...; the Boolean function ORDP (and the procedure ORDFP we defined, using it).

- ORDER: normal (i.e., ON PRI) output.

- KORDER: OFF PRI output; internal representation; recognition of common factors; REMAINDER (to be introduced on page 232); MAINVAR (to be introduced on page 215); whether the fraction 1/(A-B) prints out that way or as (-1)/(-A+B) -- that is, which variable gets the plus sign.

3.17. Domain modes

REDUCE normally uses integers as the coefficients of all polynomials and expressions built out of polynomials. We say that the normal "domain mode" is the integer number system. We can direct REDUCE to work in other domain modes:

- FLOAT -- hardware (or fixed precision) floating point

- BIGFLOAT -- arbitrary precision software floating point

- RATIONAL -- expressions like $(2*X)/3$ treated as $(2/3)*X$

- MODULAR -- coefficients integers reduced modulo some

integer.

One enters any one of these special modes, or switches from one to another, by typing the corresponding ON command: `ON FLOAT$` or `ON BIGFLOAT$` or `ON RATIONAL$` or `ON MODULAR$`. One returns to the normal domain mode by typing the corresponding OFF command. (Actually, `OFF FLOAT$` can be used in all cases.)

When changing from one domain mode to another, informative messages like

> `*** Domain mode BIGFLOAT changed to FLOAT`

or less informative messages like

> `(*** BFLOAT already loaded)`

may appear. We have suppressed these from the examples shown in this section.

These special domain modes can be used for almost any simple manipulation of polynomials and rational expressions. The more complicated operations of REDUCE, namely integration (INT), factoring (FACTORIZE and ON FACTOR), and solving equations and systems (SOLVE) may not work correctly in all cases.

The special domain modes are described more fully in the four subsections that follow.

3.17.1. FLOAT

The command `ON FLOAT$` instructs REDUCE to change numerical fractions to floating point numbers (sometimes called "real numbers"), numbers expressed to a certain limited number of significant digits and printed in decimal notation. Of course results are then subject to the roundoff errors one has learned to expect with floating point numbers.

As a simple illustration, suppose one has made a numerical

Domain modes 155

calculation using REDUCE and has discovered that the answer is

$$S := 21970635723931623979/104871010630191933$$

Accurate as this may be, it isn't very instructive. So we issue the command

ON FLOAT$

S;

209.50151

That's better. (S itself hasn't changed; we have only printed it out in a more informative way.)

The number of digits, and the exact form, differ from implementation to implementation, since the FLOAT mode uses, as much as possible, the standard floating-point hardware and software, if any, of the computer on which REDUCE is operating, and floating point format and operations are not standardized. The examples in this section were run on a DEC 20 mainframe.

If this single result is all we want of the ON FLOAT mode, we had better restore REDUCE to normal:

OFF FLOAT$

We could have typed all this on one line:

ON FLOAT$ S; OFF FLOAT$

209.50151

or, somewhat more aesthetically (as a single grouped statement rather than as three separate statements simply typed on one physical line)

<<ON FLOAT$ WRITE S$ OFF FLOAT>>$

ON FLOAT not only changes the form of printout, but the mode of the calculation itself. We present an example that shows this.

ON FLOAT$

A:=(2*X + 3*Y + 4*Z + 7*W)/3;

    A := 0.66666666*X + Y + 1.3333333*Z + 2.3333333*W

B := 3*A;

    2*X + 3*Y + 4*Z + 6.9999999*W

The coefficient which should have become 7 prints as 6.9999999 because of roundoff in the calculation. On the other hand, 3 * 0.66666666 was reported as 2, not as 1.99999998, because the result was close enough to 2 by some tolerance test. In contrast, in the example

AA := 700/7 * BB;

    AA := 100.00000*BB

the fraction 700/7, which is if course exactly 100, is reported as 100.00000, the decimal point and zeros signalling that REDUCE thought some significant roundoff had occurred (even though it hadn't).

REDUCE sometimes prints floating-point results in "E" form:

CC := DD/800;

    CC := 0.125E-2*DD

Normally (that is, in OFF FLOAT mode) if you were to enter into REDUCE an expression containing numbers with a decimal point, REDUCE would change those numbers to fractions:

Domain modes

```
C := 123.46 * X + 234.0 * Y + 789. * Z
             + 123.45 * U + 0.666 * V + 0.6666 * W;
*** 123.46000 represented by 6173/50
*** 234.0 represented by 234
*** 123.44999 represented by 2469/20
*** 0.66600000 represented by 333/500
*** 0.66660000 represented by 2/3
C := (185190*X + 351000*Y + 1183500*Z
             + 185175*U + 999*V + 1000*W)/1500
```

It may be worth while to examine these changes in representation individually.

- The first is straightforward: $123.46 = 12346/100 = 6173/50$.

- There is no difficulty with the second coefficient.

- In the third, the unnecessary point at the end was ignored without comment.

- For the next coefficient, REDUCE took the liberty of making an approximation in going from decimal form to fraction form: the **123.44999** was treated as if it were **123.45**. (Indeed it was input as **123.45**, but roundoff in the input process, combined with roundoff in output, caused it to be recorded as **123.44999**.)

- The last two coefficients show this phenomenon clearly: **0.666** was translated exactly, but **0.6666** was approximated by $2/3$. Incidentally, REDUCE would not have accepted the numbers without the zero in front of the decimal point. Floating point numbers are not allowed to <u>begin</u> with the decimal point.

These changes in representation take place both when numbers with decimal points are typed in, and when expressions created in ON FLOAT mode are evaluated in OFF FLOAT mode.

Exercise 3.17.1.

Suppose `W := Z/3;` is entered while in ON FLOAT mode, with Z clear, and then W is referred to repeatedly while in OFF FLOAT mode. What is troublesome about this? What should be done about it?

Exercise 3.17.2.

(Inspired by the above exercise) Suppose `W := Z/3;` is entered while in ON FLOAT mode, with Z clear, and then `W**3 + 5*W**2 + 7*W + 9` is evaluated while in OFF FLOAT mode. What will happen? What if we try `<<W>>**3 + 5*<<W>>**2 + 7*<<W>> + 9` (where `<< >>` should be thought of as the "superparentheses" discussed on page 112)? Does this demonstrate something about the inner workings of REDUCE?

If REDUCE is in the ON FLOAT mode, numbers with decimal points are accepted as such. Fractions in expressions are converted to floating point form.

We illustrate one more point. Observe the form of the result:

```
ON FLOAT$

D := (2*X + 3*Y)/(8*U + 9*V);

    D := (0.25*X + 0.375*Y)/(U + 1.125*V)
```

REDUCE divided through by the coefficient of one of the terms in the denominator, `8*U`. If we wanted the V in the denominator to be the variable that will have no coefficient, we can use KORDER:

```
KORDER V$

D;

    (0.22222222*X + 0.33333333*Y)/(0.88888888*U + V)
```

Domain modes

Compare this with exercise 3.16.2 on page 153.

3.17.2. BIGFLOAT and NUMVAL

In the ON FLOAT mode fractional numbers are handled as decimals with a certain standard precision, since (in most implementations of REDUCE) the computer's standard floating-point instructions are used. The command ON BIGFLOAT$ enables one to specify the precision to use, at the cost of slower operation because for BIGFLOAT software-simulated floating point is used.

```
ON BIGFLOAT$

A := 1/6;

        A := 0.166 66666 67

PRECISION 30$

A := 1/6;

        A := 0.166 66666 66666 66666 66666 66666 67
```

As this example shows, ON BIGFLOAT$ sets a floating-point mode of 10 significant digits initially. The command PRECISION N$, with any positive integer N, changes that as desired. (The PRECISION command can not be given until ON BIGFLOAT$ has been entered at least once during the REDUCE session.) Actually, BIGFLOAT works with N+2 digits internally, to reduce accumulated roundoff error, even though only N are printed in the answer. We'll refer to the extra two digits as "hidden digits".

PRECISION 0; shows the present precision value:

```
PRECISION 0;

        30
```

Once in the ON BIGFLOAT mode a second, optional, mode command is available: ON NUMVAL$. If desired the two commands can be combined: ON BIGFLOAT, NUMVAL$. The two modes can be given in either order in REDUCE 3.2. (In earlier versions of REDUCE the two had to be given in the order shown.)

With this additional command, any use of one of the standard functions like SIN or SQRT, with an argument whose value is a number, results in the function being evaluated to the current degree of precision.

SIN(A);

0.165 89613 26934 15031 89789 13555 99

SQRT(A);

0.408 24829 04638 63016 36621 40124 51

The symbols PI and E are also replaced by their numerical values to the current degree of precision.

PI;

3.141 59265 35897 93238 46264 33832 8

With only ON BIGFLOAT, SIN(A); would just echo as

SIN(0.166 66666 66666 66666 66666 66666 67)

(0.166...67 being the value of A), and PI, E would just remain as symbols.

OFF BIGFLOAT$ of course returns REDUCE to normal. If you had set NUMVAL on, you must also enter OFF NUMVAL$. (In some older versions of REDUCE, if the OFF NUMVAL$ was omitted, the BIGFLOAT mode miraculously reappeared in some circumstances even after OFF BIGFLOAT$.)

Numbers of varying precision can coexist in an expression:

Domain modes

```
ON BIGFLOAT$

PRECISION 8$

X := 2/9 * U;

        X := 0.222 22222 *U

PRECISION 16$

X := X + 5/9 * V;

        X := 0.222 22222 22*U + 0.555 55555 55555 556*V
```

Notice that, since we increased the precision, the two "hidden digits" in the expression for 2/9 are revealed. If the numbers are combined, the number of lesser precision is treated as if zeros followed the "hidden digits":

```
V := U$

X;

        0.777 77777 77555 556*U
```

If an expression created in FLOAT mode is evaluated in BIGFLOAT mode, the floating-point numbers are converted to BIGFLOAT numbers of the precision that is standard for FLOAT numbers (or less, if the current precision is less):

```
ON FLOAT$

F := 3/7 * FF;

        F := 0.42857143*FF

ON BIGFLOAT$

PRECISION 16$

B := BB / 7$

C := B + F;

        C := 0.428 57143 *FF + 0.142 85714 28571 429*BB
```

If an expression created in BIGFLOAT mode is evaluated in normal mode, BIGFLOAT numbers remain as they are. (This is in contrast to what happens to FLOAT numbers: as we saw, they get represented as fractions, integer over integer.)

OFF BIGFLOAT$

C;

   0.428 57143 *FF + 0.142 85714 28571 429*BB

We can get strange combinations:

C + 3/10*EE;

   (4.285 7143*FF + 1.428 57142 85714 29*BB + 3*EE)/10

While in normal mode we can add, subtract, and multiply BIGFLOAT numbers, but not divide them:

ON BIGFLOAT$

PRECISION 10$

U := 2/9;

   U := 0.222 22222 22

V := 3/9;

   V := 0.333 33333 33

OFF BIGFLOAT$

U+V;

   0.555 55555 56

U-V;

   - 0.111 11111 11

U*V;

0.074 07407 407

U/U;   % should be 1

$$0.222\ 22222\ 2222/0.222\ 22222\ 2222$$

Finally, we note that if an expression created in BIGFLOAT mode is evaluated in FLOAT mode, not only do the BIGFLOAT numbers keep their form and value, but any arithmetic combining FLOAT and BIGFLOAT numbers yields results in BIGFLOAT form. In fact, no matter what domain mode we may subsequently switch to, BIGFLOAT numbers remain BIGFLOAT. (This is not so with the other special number forms: for example, if we switch to the normal mode from some other mode, FLOAT and RATIONAL numbers are replaced by quotients of integers, and MODULAR numbers are replaced by integers less than the modulus.

3.17.3. RATIONAL

Normally REDUCE writes expressions -- strictly speaking, polynomials and quotients of polynomials -- with at most a single division "/" operation. If the coefficients of the polynomials include fractions, REDUCE "multiplies through" to clear fractions. For example,

S := FOR I:=0:4 SUM X**I/(I+3)**2;

$$S := (3600*X^4 + 4900*X^3 + 7056*X^2 + 11025*X + 19600)/176400$$

Each of the five terms that were added had a denominator (9, 16, 25, and so on). The sum was then automatically rewritten with a single denominator, 176400, totally obscuring the structure of the expression.

The command ON RATIONAL$ prevents <u>numeric</u> denominators from being combined (or, if they had been combined, separates them).

ON RATIONAL$

S;

$$1/49*X^4 + 1/36*X^3 + 1/25*X^2 + 1/16*X + 1/9$$

This mode doesn't <u>always</u> make expressions easier to understand.

```
W := (2*X + 3*Y)/(5*X + 7*Y);

W := (2/5*X + 3/5*Y)/(X + 7/5*Y)
```

Instead of keeping the simple integer coefficients in the numerator and denominator of W, computation in the ON RATIONAL mode put W in a "standard" form in which the coefficient of one of the terms in the denominator was forced to be 1, by dividing the numerator and denominator through by 5.

We can use **KORDER** to get Y to be the variable whose coefficient is 1:

KORDER Y$

W;

$$(2/7*X + 3/7*Y)/(5/7*X + Y)$$

We return our attention to the sum S. The presence of a denominator with its own coefficients again covers up the structure of the sum:

```
S/(5*A + 7*B);
```

$$(1/245*X^4 + 1/180*X^3 + 1/125*X^2 + 1/80*X + 1/45)$$
$$/(A + 7/5*B)$$

It may be useful to observe that the fractional coefficients correspond exactly to the floating point coefficients one gets in ON FLOAT mode:

ON FLOAT$

WS;

$$(0.40816326E\text{-}2*X^4 + 0.55555555E\text{-}2*X^3 + 0.80000001E\text{-}2*X^2 + 0.125E\text{-}1*X + 0.22222222E\text{-}1)/(A + 1.3999999*B)$$

In summary, the REDUCE user should have this mode in his armory of techniques, but not expect more from it than it is capable of giving.

3.17.4. MODULAR

Modular arithmetic is primarily of interest to students and researchers in number theory, abstract algebra, and some other areas of pure mathematics. Others can omit this subsection without loss.

REDUCE can be directed to carry out its computations modulo M, where M is any positive integer. This means that any coefficient equal to or larger than M, in any polynomial, is replaced by the remainder resulting from dividing the coefficient by M. For example, if M is 12 then 10+10 is not 20 but 8.

Some computations do not work if M is not a prime. Certain cases of this should be expected, but there are others in which there is a correct answer but REDUCE can't obtain it.

It takes two steps to activate modular arithmetic:

SETMOD 5$

ON MODULAR$

These two steps should not be interchanged: the SETMOD command, through which the user supplies the value of M, should be given while REDUCE is still in its normal domain mode.

Now after the ON MODULAR$ command, modular arithmetic is in effect. Examples:

```
A := (X**2 + Y**2)**4;
```

$$A := X^8 + 4*X^6*Y^2 + X^4*Y^4 + 4*X^2*Y^6 + Y^8$$

Notice that the middle coefficient, which would normally be 6, has been reduced mod 5 to 1. Also notice that the modular reduction applies only to the coefficients. The exponents remain unreduced integers.

```
A := (X**2 + Y**2)**5;
```

$$A := X^{10} + Y^{10}$$

The correct answer.

There are some other contexts, besides exponents, in which numbers are allowed unreduced. The limit specified in a FOR statement is not reduced:

```
FOR I := 0:10 SUM X**I;
```

$$X^{10} + X^9 + X^8 + X^7 + X^6 + X^5 + X^4 + X^3 + X^2 + X + 1$$

If the summation limit had been reduced mod 5, only a single term (with I=0, as if we had said FOR I := 0:0) would have been obtained.

If M is a prime, division by numbers not divisible by M causes no problems. Still using M=5, we get

```
3/7;
                4
4*7;            % check
                3
```

Domain modes

Some more examples:

```
(A + 2*B + 3*C + 4*D)/7;

        3*A + B + 4*C + 2*D
WS*7;              % check
        A + 2*B + 3*C + 4*D
1/(3*A + 4*B);

        2/(A + 3*B)
```

In the last example, the numerator and denominator were divided by 3 (i.e., multiplied by 2) to make the coefficient of one of the variables in the denominator equal to 1.

Now let us try a value for M that is not a prime.

```
OFF MODULAR$      % before resetting the modulus
SETMOD 10$
ON MODULAR$
3/7;

        9

3/5;

        ***** Invalid modular division
```

Since 5 and the modulus M=10 have a factor in common, we can't divide by 5.

```
1/(3*X + 5*Y);

        7/(X + 5*Y)
```

Now suppose the internal ordering of X and Y were reversed:

```
KORDER Y,X$

A;

        ***** Invalid modular division
```

What happened was that REDUCE now considered Y to be the "first" variable in the denominator, and tried to make its coefficient equal to 1. But that requires the impossible operation of division by 5!

(We could have illustrated the same phenomenon without using KORDER, by attempting to evaluate 1/(5*X + 3*Y).)

The division 1/(5*X + 2*Y), with modulus M=10, can't be carried out by REDUCE no matter what KORDER is used, because both 5 and 2 have factors in common with 10.

We point out that if all that is required is modular reduction of a few expressions that are polynomials (i.e., contain no fractions), it may be simpler to use the REMAINDER function than putting REDUCE into MODULAR mode. A single example should suffice:

A := (X**2 + Y**2)**4$

B := REMAINDER(A,5);

$$B := X^8 + 4*X^6*Y^2 + X^4*Y^4 + 4*X^2*Y^6 + Y^8$$

This is the same as the answer obtained using **SETMOD 5$** and **ON MODULAR$** earlier in this subsection.

(REMAINDER and the MODULAR mode treat negative numbers differently. Under MODULAR mode, with modulus 5, the number -7 becomes +3. The REMAINDER(-7,5) is -2.)

4. Procedures

Most programming languages allow users to package frequently used sequences of statements in an easy-to-use way. These packages are called "procedures" (or "functions" or "subroutines"). REDUCE is no exception. However, REDUCE is so powerful a tool even when used in "calculator mode" (i.e. by typing in commands and examining the results obtained) that user-defined procedures play a much lesser role in it than in most other languages.

Procedures for REDUCE can be written either in REDUCE or in the underlying language RLISP. Features of REDUCE such as NUM, DEN, DF, INT, SUB, and PART are system procedures written in RLISP that manipulate the internal representation of expressions. We will not attempt to explain these internal forms or how to program in RLISP. The reader should be aware, though, that if he knew these things he could write procedures that generally execute much faster than those written in REDUCE (but at a far greater cost in programming and debugging time).

The definitions of procedures written in RLISP begin with the words SYMBOLIC PROCEDURE, or, equivalently, LISP PROCEDURE. The definitions of procedures written in REDUCE begin with the word PROCEDURE, or if preferred for emphasis ALGEBRAIC PROCEDURE. We will be concerned exclusively with (ALGEBRAIC) PROCEDUREs.

In order to allow knowledgable users relatively easy access to the whole REDUCE source program, system procedures are not protected against user redefinition. Whenever any procedure is redefined, a message

    *** <procedure name> REDEFINED

is printed. If this occurs, and the user is not redefining his own procedure, he is well advised to rename it, and possibly start over by exiting from REDUCE and calling in a fresh copy of REDUCE just as he did at the beginning of the session, because he has already redefined some internal procedure whose correct functioning may be required for his job!

4.1. Procedures without parameters or RETURN

The simplest procedures are those whose exact function is fixed for once and for all. For example, suppose that during a lengthy series of computations the user frequently wants to check on the values of the variables APPLE, BOOK, CAT, and DOG. He could type a series of commands APPLE; BOOK; CAT; DOG; each time, or he could for once and for all define

```
PROCEDURE ABC;
BEGIN
    WRITE "APPLE = ",APPLE;
    WRITE "BOOK = ",BOOK;
    WRITE "CAT = ",CAT;
    WRITE "DOG = ",DOG;
END;
```

(The semicolon before the END, that is, after DOG, is not really necessary, but is harmless.) Then whenever he wants to see the values of the four variables, each neatly identified, he merely calls on the procedure by typing

ABC()$

to which REDUCE reponds with, perhaps,

APPLE = RED

BOOK = PAPERBACK

CAT = SCRATCH + MEEOW

DOG = BARK*BITE

The procedure is defined by simply typing it in as shown above. Just as entering PQR:=... assigns to PQR a "value" that is recorded in the system, typing

PROCEDURE ABC; BEGIN ... END;

(arranged in one line or on as many lines as is convenient and

Procedures without parameters or RETURN 171

readable) attaches to ABC a "procedure definition" that is recorded in the environment. Just as the "value" of PQR can be changed by entering a new assignment, the "procedure definition" of ABC can be changed by entering a new definition. CLEAR ABC does not erase the procedure definition.

To use the procedure once the definition has been stored, the name is typed, followed by an empty pair of parentheses () and then a dollar sign terminator. The statements that were listed between the BEGIN and the END are then carried out. (If a semicolon terminator is used, a meaningless zero prints below the desired output.) The parenthesis pair distinguishes a call on a procedure from the evaluation of a variable. If ABC; were typed (with no parentheses, and semicolon as terminator) the "value" of ABC would print, normally ABC if ABC is clear, i.e. if no value assignment (through := or LET) has also been made for ABC. If ABC$ were typed there would be no output at all.

In defining a procedure it makes no difference which of the two terminators (semicolon or dollar sign) is used between the statements. Most programmers find the semicolon to be the easier to type and read.

There is an alternative to defining ABC as a procedure. We can use LET:

```
LET ABC =
  BEGIN WRITE "APPLE = ",APPLE;
        WRITE "BOOK = ",BOOK;
        WRITE "CAT = ",CAT;
        WRITE "DOG = ",DOG;
  END;
```

Since this is a LET instead of a ":=" assignment, the right-hand side of the equal sign is stored unevaluated -- essentially verbatim -- as the "value" of ABC. To use the ABC so defined, type in ABC$, with no parentheses. The "value" of ABC is then "computed". The actual result computed is a meaningless zero that is not printed since the terminator is a dollar sign, but in the course of the "computation" the four WRITE statements are executed.

We give another example showing the potential usefulness of "parameterless procedures" such as we are discussing. Suppose that during the same lengthy calculation the variable A receives values from time to time, and we want to create a "historical record" of certain of these values. On demand, we want to be able to store the current value of A in the next available position of an array.

The first step is to declare an array, and set a variable to zero indicate that, so far, the array is empty:

```
ARRAY SA(100)$
SI := 0$
```

Now to define a convenient procedure:

```
PROCEDURE SAVEA;
BEGIN
    SA(SI) := A;
    SI := SI + 1;
END;
```

Whenever the command SAVEA()$ is entered, the current value of A is stored in the array and the array index SI is updated. (Of course SI should be left otherwise undisturbed.)

Again, a LET could be used instead:

```
LET SAVEA =
  BEGIN
    SA(SI) := A;
    SI := SI + 1;
  END;
```

To use this, enter SAVEA$, without a parenthesis pair.

We note that for all the examples in this section, the grouping symbols << and >> could have been used instead of BEGIN and END. This won't be the case when procedures with local ("SCALAR") variables are being defined.

Procedures without parameters or RETURN   173

Exercise 4.1.1.

Write and test a procedure SABC that sets the variable S3 equal to A+B+C, and also prints out a message

    THE SUM IS NOW ...

displaying the value of S3. (Note: much as we might have liked to name either the variable or the procedure SUM, we couldn't do so because SUM is a reserved word, used in the FOR ... SUM construction.)

Exercise 4.1.2.

Define SABC as a LET statement instead.

During a REDUCE session it may be necessary to input some particular LET rule, and subsequently CLEAR it, several times. One use of procedures without parameters is to package such a rule and its corresponding CLEAR. Groups of rules can be handled equally well. The user might, for example, define

```
PROCEDURE L1;
BEGIN
        LET X**5=0;
        LET U*V=SSS;
        FOR ALL X LET H(X)=0;
END;

PROCEDURE C1;
BEGIN
        CLEAR X**5;
        CLEAR U*V;
        FOR ALL X CLEAR H(X);
END;
```

Suppose we had

    W := X**7 + 2*U*V + H(P+Q)$

Calling on L1 puts the LET rules into the substitution environment:

L1()$

W;

    2*SSS

Calling on C1 deletes the LET rules from the substitution environment:

C1()$

W;

$$H(P + Q) + X^7 + 2*U*V$$

(We picked the names L1 and C1 with the thought that other groups of LET rules and corresponding CLEAR's might be packaged as L2 and C2, etc.)

Once again, LET rules could have been used instead (one LET rule placing other LET rules into the substitution environment!):

```
LET L1 =
     <<LET X**5=0;
       LET U*V=SSS;
       FOR ALL X LET H(X)=0;>>;

LET C1 =
     <<CLEAR X**5;
       CLEAR U*V;
       FOR ALL X CLEAR H(X);>>;
```

Now L1$ will place the package of LET's into the substitution environment, and C1$ will clear them.

4.2. Procedures with RETURN

The procedures discussed in the preceding section performed actions. They didn't return values. For example, if we called SAVEA using SAVEA(); instead of SAVEA()$, the value printed out would be meaningless and irrelevant to the job SAVEA was supposed to do. Similarly, if we had written X := SAVEA()$, the value saved

Procedures with RETURN 175

in X would be meaningless. Using the terminology of many programming languages, the procedures acted like subroutines, not functions.

By "returning" a result, we can have our procedures serve as functions. Recall that our SAVEA procedure stored data in an array SA, using the variable SI to keep track of where in SA the next item should be put. Suppose we frequently need the average of the two most recent items in the SA array. We define a procedure for the purpose:

```
PROCEDURE AV2A;
BEGIN
        RETURN (SA(SI-1) + SA(SI-2))/2
END;
```

Now XX := AV2A()$ would set XX equal to the desired average; AV2A(); would automatically print out the desired average; and so on. A call AV2A()$ would have no visible result, because the calculated result is thrown away. (Not completely thrown away: WS holds the value, as always.)

If SAVEA has been called only once, there is no SA(SI-2) to use in the average. Indeed we'd get an OUT OF RANGE message informing us that we tried to use a negative index (SI-2) = -1 in the reference to array SA. Let us rewrite AV2A so that in that case it doesn't attempt to form an average:

```
PROCEDURE AV2A;
BEGIN
        IF SI > 1 THEN
            RETURN (SA(SI-1) + SA(SI-2))/2;
        RETURN SA(SI-1)
END;
```

This illustrates two points.

1. The word RETURN can occur more than once in a procedure. It doesn't have to be in the last line of the procedure.

2. Once a RETURN is executed the rest of the procedure is

not obeyed. If SI > 1 then the first RETURN is obeyed; the second is ignored. (We could replace the semicolon at the end of the first RETURN by an ELSE with no change in effect.)

Exercise 4.2.1.

Revise AV2A once more so that if SI is zero when AV2A is called then zero is returned as result but first a message

*** NO A VALUES SAVED YET ***

is printed out.

RETURN returns the entire expression between the word RETURN and the next terminator or END. Thus RETURN X+Y can be written, with no parentheses required. RETURN; (with no expression given) is equivalent to RETURN 0;. The word RETURN is permitted only within BEGIN ... END constructions. If the grouping construction << ... >> is preferred, the value to be returned must be placed last in the angle brackets, being careful not to write a terminator between it and the closing >>:

```
PROCEDURE AV2A;
    <<IF SI > 1 THEN
        (SA(SI-1) + SA(SI-2))/2
     ELSE SA(SI-1)>>;
```

(Here the expression desired is not only the last in the brackets, it is the only one. What's is the brackets is a single conditional -- IF ... THEN ... ELSE -- expression.)

Exercise 4.2.2.

Do the last exercise over using << ... >> instead of BEGIN ... RETURN ... END.

If the desired result is given by a single expression, as was the case in the example above, the grouping brackets << ... >> aren't necessary.

Procedures with RETURN   177

Exercise 4.2.3.

Do the last exercise over not even using << ... >>.

Some do's and don'ts about where RETURN can be used follow. They can be ignored at first reading.

RETURN can be used within a << ... >> construction within the BEGIN ... END procedure body.

RETURN can be used within an IF ... THEN ... ELSE ... within the BEGIN ... END procedure body.

RETURN must not be used within a FOR, WHILE, or REPEAT statement within the procedure body.

Note: There is a bug in some versions of REDUCE that, in very special circumstances, causes RETURN to return the wrong value. To see if this is the case with the version you are using, enter the following:

```
PROCEDURE AAA F;
BEGIN
G := F;
RETURN SUB(X=5,G)-SUB(X=3,G);
END;

AAA(X);
```

The answer should, of course, be 2. If you get the answer 0, you have a flawed version of REDUCE. But you can prevent the flaw from possibly causing an error very simply: always write "WS:=" between RETURN and the expression to be returned. (This can't do harm in any case.) In this example, changing the RETURN line to

```
RETURN WS :=   SUB(X=5,G)-SUB(X=3,G);
```

will cause the response to the AAA(X) call to be 2.

4.3. Returning multiple values

We could summarize this section, Returning Multiple Values, by saying "It is not possible". RETURN returns only a single expression. The purpose of this section is to suggest techniques for getting around this restriction. (In future versions of REDUCE there may be more direct solutions.)

1.

If the procedure is intended to return a fixed small number of expressions, say three results, it could assign the three values to three specially designated variables, say ANS1, ANS2, and ZZZ:

```
PROCEDURE ABC;
BEGIN
            . . . . . . . .
            . . . . . . . .
            ANS1 := ... ;
            ANS2 := ... ;
            ZZZ  := ... ;
END;
```

This procedure would be used by executing it, and then "picking up" the values of ABC1, ABC2, ZZZ as required. For example if all three need to be printed the appropriate input would be

```
ABC()$
ANS1;
ANS2;
ZZZ;
```

or all on one line:

```
ABC()$ANS1;ANS2;ZZZ;
```

Of course when this procedure is going to be used the three variables ANS1, ANS2, ZZZ had better not have prior values that are valuable, because the call ABC()$ will have the "side effect" of destroying those values.

Notice that the ABC() is followed by a dollar sign and not a semicolon, because ABC as written above doesn't RETURN anything.

Returning multiple values

We could eliminate one of the three variables by RETURNing one of the three results:

```
PROCEDURE ABC;
BEGIN
        ........
        ........
        ANS2 := ...;
        ZZZ  := ...;
        RETURN ...;   % what went to ANS1 before
END;
```

If this is the only change made,

```
ABC();ANS2;ZZZ;
```

would print the same three results in the same order as before.

2.

If there is a larger but fixed (or at any rate bounded) number of results the procedure is intended to produce, the procedure could put them in a specially designated array.

```
PROCEDURE ABC;
BEGIN
        ........
        ........
        FOR I:=1:100 DO
            ANSS(I) := ...;
END;
```

This presupposes that ANSS has been declared to be an array (of size at least 100) before the procedure ABC is ever called.

The declaration **ARRAY ANSS 100;** can be placed inside the procedure definition, but to prevent the annoying message

```
*** ANSS ALREADY DEFINED AS ARRAY
```

every time (except the first time) the procedure is called, one must precede the declaration by **CLEAR ANSS;**:

```
PROCEDURE ABC;
BEGIN
        CLEAR ANSS;
        ARRAY ANSS 100;
        . . . . . . . .
        . . . . . . . .
        FOR I:=1:100 DO
            ANSS(I) := ...;
END;
```

Obviously neither the array ANSS nor the ordinary variable ANSS should contain useful information before ABC is called, because the call will destroy that information.

3.

In the first approach, we had to "reserve" several specific variable names to receive the results. In the second approach we only had to "reserve" a single name, to be the name of an array holding the results. We now describe another approach, one in which again a single variable is committed to the aim of somehow RETURNing more than one value, but now the only requirement on this variable is that it be clear initially -- and it will remain clear.

Let us suppose that the variable AA is sure to be clear, and sure not to occur in any of the results to be returned. Then we can "pack" the expressions, say the three expressions 95, (X+Y), and SIN(Z), into a single expression by forming

   95*AA + (X+Y)*AA**2 + SIN(Z)*AA**3

To use this idea, we rewrite our procedure so as to construct and return this kind of expression:

```
PROCEDURE ABC;
BEGIN
        . . . . . . . .
        . . . . . . . .
        RETURN FOR I:=1:100 SUM
            (...) * AA**I;
END;
```

To use this version of the procedure, we might write

Returning multiple values 181

```
V := ABC()$
```

Then we can use COEFF to unpack V to get the individual results needed.

The disadvantage of this approach is the necessity of unpacking the result. But an important advantage is that V can be copied, saved, compared, etc. all at once -- something we can't do with an array. For example, if we need to check whether ABC computes the same 100 results at two different times during a REDUCE session, we could use

```
V := ABC()$
........
........
ABC() - V;
```

If the difference in the last line prints out as 0, we know there is no change. If a term involving AA**37 prints out, we know that the 37'th result has changed.

4.

We can also use essentially the same idea with several clear variables. Suppose (referring back to the first approach we presented) ANS1, ANS2, and ZZZ are guaranteed to be clear, and guaranteed not to occur in any of the results to be returned. Instead of having the procedure assign values to these three variables, we could use them as term-markers in a single RETURNed expression:

```
PROCEDURE ABC;
BEGIN
        ........
        ........
        RETURN ANS1 * (...)
             + ANS2 * (...)
             + ZZZ  * (...);
END;
```

This version of ABC could be used by calling

```
V := ABC()$
```

To print the three results, we could unpack them in any of several ways, perhaps the simplest of which is to use differentiation:

```
DF(V,ANS1); DF(V,ANS2); DF(V,ZZZ);
```

Again, this mechanism gives us a simple way to save, compare, etc. all three results simultaneously.

4.4. Procedures with one parameter

So far all our procedures worked with the same variables each time they were called. For example, SAVEA always saved the value of A, and could not be used to save the value of B except by writing A:=B$ before calling SAVEA()$.

We now define a more flexible version of SAVEA, called SAVE:

```
PROCEDURE SAVE(X);
BEGIN
    SA(SI) := X;
    SI := SI + 1;
END;
```

To use this to save the value of A in array SA, enter SAVE(A)$. To use this to save the value of B in array SA, enter SAVE(B)$.

What happens when you enter the command SAVE(B)$ is that the value of B is automatically copied to the variable X named in the parentheses following the procedure name in the procedure definition. Then the statement SA(SI) := X in the procedure copies the value of X (which is also the value of B) into the SA array.

The actual parameter (such as B) does not even have to be a variable. It can be any expression. SAVE(P+Q)$ saves the current value of the sum P+Q in the SA array.

If the actual parameter is just a variable, or a number without a minus sign, the parentheses in the call are not necessary. SAVE B$ can be used instead of SAVE(B)$. This is just the same notational convenience as is provided for referring to array places and operator

Procedures with one parameter

"places". Similarly, the parentheses are optional in the procedure heading: we could have written **PROCEDURE SAVE X;**.

The variable X in the procedure heading is referred to as a formal parameter, because it is used to show the form of the action to be taken. Formal parameters are local to the procedure. What this means is that the formal parameter, here X, is distinct from the variable X that may be in use elsewhere in the program. If X was clear before the command **SAVE(B)$** or **SAVE(P+Q)$** was issued it will be clear after the command is obeyed. If X was equal to 1357 before, then it will be equal to 1357 after. This point will be discussed at greater length in section 4.8 on Linkage Questions.

Exercise 4.4.1.

Test the statements about formal parameters that were just made. In the test use a procedure, with formal parameter X, that contains within it a statement **WRITE "INSIDE, X = ",X;**.

No assignment statements that change the value of the formal parameter should be written inside the procedure. For example, suppose a procedure requires only the numerator of the actual parameter. If X is the formal parameter, one is tempted to begin the procedure body with **X := NUM X**. He should instead write **XX := NUM X**, or the like, and work with XX in the remainder of the procedure. Here XX should be a "local variable", a concept that is explained a few pages hence. The reasons for this warning are too technical to explain here. Suffice it to say that in some cases it leads to failure; in most cases the assignment seems to do what was intended, but, because of the way formal parameter values are represented, it makes the procedure inefficient (i.e., run slowly -- sometimes very slowly). But since making assignments to formal parameters doesn't usually cause noticeable trouble, this warning is often ignored.

Any symbol, not just X, can be used as the formal parameter, with two exceptions: T and NIL. The exceptions are a consequence of the special roles T and NIL play in RLISP, the language in which REDUCE is defined. (The user is not really likely to try to use the symbol NIL for a formal parameter, but unfortunately one is often tempted to name a formal parameter "T"!)

We finish this section with an example of what might be considered to be the simplest kind of procedure: a function that takes one value, and from it computes a result using a straight-forward formula.

If $F(X)$ is to mean $(X+5)/(X+6)$, so that $F(10)$ would evaluate to $15/16$, $F(A-6)$ to $(A-1)/A$, and so on, the entire procedure defining F could be

```
PROCEDURE F X;
    (X+5)/(X+6);
```

The definition could even be written on a single line. Since the body of the procedure is a single statement, no << ... >> or BEGIN ... RETURN ... END is needed.

Exercise 4.4.2.

Define and test a function $F(X)$ of your own choosing.

Exercise 4.4.3.

Define a procedure FLOT such that if A is an expression, the call **FLOT A$** will print out A converted to the ON FLOAT form. It will accomplish this by using a WRITE statement located inside the body of the definition. (Something rather similar was accomplished on page 155.) FLOT is primarily intended for the sake of this "side effect" of printing something, not to act like a function and return an answer. However, if the "result" of the procedure is captured, as by calling it from a statement like **B := FLOT(A)$**, this result should be A unchanged.

The procedure FLOT should assume that REDUCE is in its usual OFF FLOAT mode when it starts, and should leave it in that mode when done.

4.5. Procedures with more than one parameter

A procedure can have more than one formal parameter. If the procedure needs to be given the values of several variables or expressions, list the appropriate number of formal parameters in parentheses in the heading, separated by commas:

```
PROCEDURE ABC (X,Y,Z);
BEGIN ... END;
```

If there is more than one formal parameter, the parentheses are required in the heading.

As an example, suppose that during a computation with polynomials involving the two variables X and Y we frequently need to find the coefficient of the X∗Y term. The reader who knows Calculus will recognize that if A is the name of the polynomial then SUB(X=0,Y=0,DF(A,X,Y)) will give us this coefficient. (Recall that DF(A,X,Y) means $d^2A/dXdY$.) We first make this into a procedure with one parameter:

```
PROCEDURE CXY A;
BEGIN
    RETURN
        SUB(X=0,Y=0,DF(A,X,Y))
END;
```

(Since our "usual" formal parameter symbol X is needed for other purposes, we picked A for formal parameter this time.)

If we wanted the procedure to be able to handle variables other than X and Y, we could add them to the parameter list:

```
PROCEDURE CXY(A,X,Y);
BEGIN
    RETURN
        SUB(X=0,Y=0,DF(A,X,Y))
END;
```

To use this version of CXY to find the coefficient of X1∗X2 in the value of Q, one would write

```
CXY(Q,X1,X2);
```

(The variables X1 and X2 had better be clear, or at least equal to variables and not more complicated expressions, or the DF operation will complain!)

Exercise 4.5.1.

Write a procedure, using a single SUB operation, that returns the result of interchanging any two specified variables appearing in any specified expression.

We now present a mathematically more interesting example. Suppose we need the so-called Legendre polynomials in our work. The textbook formula for the Legendre polynomial of order n is

$$P_n(x) = \frac{1}{n!} \frac{d^n}{dy^n} \frac{1}{(y^2-2xy+1)^{1/2}} \Bigg]_{y=0}.$$

We want a function P(N,X) that, for any positive integer N, is the Legendre polynomial of order N. Put into words, P(N,X) is the result of substituting Y=0 in the N'th partial derivative with respect to Y of a certain fraction involving X and Y, then dividing that by the factorial of N.

This verbal formula can easily be written as a single expression in REDUCE:

```
PROCEDURE P(N,X);
    SUB(Y=0,DF(1/(Y**2 - 2*X*Y + 1)**(1/2),Y,N))
         /(FOR I:=1:N PRODUCT I);
```

Having input this definition, we can use it in any formula in which it may be needed:

```
A := 20*P(2,W);

              2
    A := 30*W  - 10
```

Let us summarize:

Procedures with more than one parameter

Each procedure has a heading consisting of the word PROCEDURE (optionally preceded by the word ALGEBRAIC), followed by the name by which the procedure is to be referred to, and followed by its formal parameters -- the symbols that will be used in the body of the definition to illustrate what is to be done. There are three cases:

1) No parameters. Simply follow the procedure name by a terminator (semicolon or dollar sign).

When such a procedure is used in an expression or command, ABC(), with empty parentheses, must be written.

2) One parameter. Enclose it in parentheses (or just leave at least one space), then follow by a terminator.

    PROCEDURE ABC(X);
or **PROCEDURE ABC X;**

3) More than one parameter. Enclose them in parentheses, separated by commas, then follow by a terminator.

    PROCEDURE ABC(X,Y,Z);

To continue our summary, there are also three types of procedure bodies:

1) A single expression or formula whose value is to be the value of the procedure.

2) A series of statements, the value of the last of which is to be the value of the procedure. The series is enclosed in << ... >> grouping symbols, and the statements are separated by semicolons (or, equivalently, dollar signs). There must not be a terminator after the last statement; if there is, zero will be returned as the answer regardless of what calculations were made.

3) A series of statements, enclosed in BEGIN ... END and separated by semicolons (or, equivalently, dollar signs). When a

statement of the form RETURN ... is executed, the computation ends with the indicated value as the value of the procedure. This form, while more awkward than the second, is necessary under two conditions: first, if there is more than a single point from which RETURN could be required, or that point can't naturally be at the end; second, if "local variables", as described in the next section, are necessary.

Exercise 4.5.2.

Write a collection of very simple procedures for manipulating what the user sees as equations. We want

- a two-parameter procedure MAKE such that a call like MAKE(X**2 + B*X + C, Y**3)$ will print out

$$X^2 + X*B + C = Y^3$$

    and also set a global variable named LT equal to the expression on the left side of the printed equal sign, and another global variable RT equal to the expression on the right. (By "global variable" is meant, for now, an ordinary variable that is not a formal parameter nor declared special in any other way.)

- a one parameter procedure AQ such that AQ(W)$ adds the expression W to both sides of the equation (that is, to the variables LT and RT), and prints the resulting equation.

- similar one parameter procedures SQ, MQ, DQ to subtract an expression from both sides or to multiply or divide both sides by an expression.

Note that all the procedure calls end with a dollar sign, not a semicolon; the printing of the equations must come from a WRITE statement inside each procedure. No RETURN is needed in any of the procedures. If one of the procedures is erroneously called using a semicolon, a meaningless 0 could be expected to print out below the equation.

Procedures with more than one parameter

To make this exercise clearer, we show below a short example of how such a package of procedures could be used:

U:=3*A + W*B + A*W$

FACTOR A,B$

MAKE(U,O)$

      A*(W + 3) + B*W = O

SQ(B*W)$

      A*(W + 3) = - B*W

DQ(W+3)$

      A = (- B*W)/(W + 3)

A:=RT;   % to give A the value just discovered

      A := (- B*W)/(W + 3)

U;   % test that it really makes U zero

    O   % yes!

4.6. Procedures with local variables

If intermediate results must be formed and saved temporarily during the computation a procedure is performing, local variables must be provided in which to store them. "Local" means that their values are deleted as soon as the procedure's operations are complete, and that there is no conflict with variables outside the procedure that may happen to have the same name.

Procedures that are to have local variables must be of the BEGIN ... END form, not << ... >>.

Local variables are created by a SCALAR declaration that must be placed immediately after the BEGIN:

SCALAR A,B,C,Z;

If more convenient, several SCALAR declarations can be given one after another:

    SCALAR A,B,C;
    SCALAR Z;

(In place of SCALAR the reader may occasionally find the declarations INTEGER or REAL in published REDUCE procedures. In the present version of REDUCE there is no difference between the three, and so no reason to use anything but SCALAR.)

Caution: The SCALAR declaration(s) can only be given immediately after a BEGIN. SCALAR must not, for example, be used on the "top level", as an interactive command to REDUCE.

All variables declared SCALAR are automatically initialized to zero.

The symbols T and NIL can not be used as local variables, for the same reason that they can not be used as formal parameters.

Local variables can not be used as arrays, matrices, or operators, only as "ordinary" REDUCE variables. If arrays, matrices, or operators are needed for a procedure's internal operation, one must pick names for them that are not likely to be used anywhere else, and declare them ARRAY or MATRIX or OPERATOR before defining the procedure. (Putting the declarations in the procedure itself leads to annoying messages "... **ALREADY DEFINED AS** ..." or "... **REDEFINED**" every time after the first time the procedure is used.)

Any symbols in the procedure body that are not formal parameters (i.e. not listed in the procedure heading), and not declared local by a SCALAR declaration, are global, in the sense that they refer to the variables of the same name outside the procedure. In particular, their values are permanently changed if assignments are made to them.

To illustrate, here is the Legendre Polynomial example of page 185 rewritten as a series of steps instead of a single formula:

Procedures with local variables

```
PROCEDURE P(N,X);
BEGIN
        SCALAR SEED,DERIV,TOP,FACT;
        SEED:=1/(Y**2 - 2*X*Y +1)**(1/2);
        DERIV:=DF(SEED,Y,N);
        TOP:=SUB(Y=0,DERIV);
        FACT:=FOR I:=1:N PRODUCT I;
        RETURN TOP/FACT;
END;
```

As a mathematically interesting example of a different kind, consider the following procedure NID that will change a fraction with the complex number i appearing in the denominator to an equivalent one that has "No i in the Denominator" (hence the name NID). Recall that REDUCE has a predefined LET I**2 = -1 rule so that any I that is not local to a procedure (or to a FOR statement or FOR expression) acts like the square root of minus 1.

```
PROCEDURE NID A;
BEGIN
        SCALAR P,Q,R,S;
        P:=NUM A; Q:=DEN A;
        R:=SUB(I=-I, Q);
        P:=P*R; Q:=Q*R;
            ON GCD;
            S:=P/Q;
            OFF GCD;
        RETURN S
END;
```

This uses the technique taught in high school algebra: to multiply both the numerator and denominator by the complex conjugate (here called R) of the denominator.

It would be awkward, but not impossible, to write the above without using at least some local variables. We used four for clarity -- an aim that is frequently more important than some abstract goal of economy.

Note the defect, almost unavoidable without descending to the RLISP level, that if the user had previously set the GCD mode to ON this procedure turns it off without warning.

Exercise 4.6.1.

Define some simple procedure with one parameter X and one local variable ABC. Verify that after using this procedure both X and ABC have the values (or clear state) they had before the use. Verify that there is no difficulty even if the actual parameter happens to be called X or ABC, or is an expression involving X and ABC.

4.7. Interaction of procedures

The only kind of statement that can not appear inside the defining body of a procedure is the definition of another procedure. All procedures are defined "on the same level", in contrast to some other programming languages such as ALGOL and PASCAL. The normal environment in which the REDUCE user works is one in which he has defined a number of procedures, to all of which he then has equal access.

While one can not define a procedure within a procedure, one can most certainly use procedures within procedures. Inside the defining body of a procedure ABC we can include statements making use of a procedure PQR. Before using (i.e., calling) ABC we must obviously have defined ABC and also PQR, but it doesn't matter which we define first: ABC or PQR.

During a REDUCE session we can define procedures and later give them corrected or entirely different definitions. Suppose we define PQR, define ABC (which calls on PQR), redefine PQR, and then call on ABC. REDUCE will use the new definition of PQR while carrying out the steps specified in the definition of ABC, not the definition that was in the system at the time ABC was being defined.

If we had forgotten to define PQR, and we attempt to use ABC, we get a message

    DECLARE PQR OPERATOR ?

We have two choices:

Interaction of procedures   193

- Reply N, hit carriage return, type in the definition of PQR, and try again. "Try again" means again typing in the command that resulted in the DECLARE question; or typing in RETRY; as explained in the chapter on Running REDUCE; or, if your version of REDUCE provides numbered prompts for the commands entered and the offending command was line 93, typing in "INPUT 93;". This is the safest choice, but if it means that a great deal of effort (human or computer) will have been wasted one might try another remedy:

- Reply Y and hit carriage return. Assuming that ABC is going to use the result of the PQR calculation only to form part of the answer expression, and not in a test like IF NUMBERP PQR(...) THEN ... or in some other subtle operation, the answer to ABC will contain the operator symbol PQR(...). Assume that the result has been saved, say as the value of the variable ZZZ. Type in CLEAR PQR$ to un-declare that PQR is an operator; type in the definition of the procedure PQR; and then ask for the value of ZZZ. The procedure definition will cause the PQR(...) symbol in ZZZ to be evaluated. (This is the only alternative available if the command using ABC was read from a file, since in that mode PQR is automatically declared OPERATOR, with a message issued to that effect, and you have no opportunity to reply N.)

A procedure ABC can even call itself, ABC, to accomplish some preliminary step. Obviously this so-called recursive call can not be required in every situation in which ABC is called, for then we would have an endless process of ABC calling ABC calling ABC ... . Eventually ABC has to be called with arguments that do not require a subsidiary ABC calculation for determining the answer. When recursion is in the picture REDUCE has no trouble distinguishing between the copies of the formal parameters and SCALAR variables at the several levels.

We give the standard "textbook example" of recursion, a recursive procedure to calculate the factorial of any positive integer.

(A procedure using FOR ... PRODUCT would be simpler and more efficient.)

```
PROCEDURE FACT N;
IF N=1 THEN N
       ELSE N*FACT(N-1);
```

4.8. Linkage questions

4.8.1. Formal parameters

4.8.1.1. The copy variable

If the symbol X is a formal parameter, the value of the corresponding actual parameter is automatically copied to a variable X "belonging to" the procedure. To emphasize that this is not the same as the variable X of the user's program, let us call it MX if the name of the procedure happens to be M. (Of course if the user's program contains a variable actually named MX we aren't talking about that variable either.)

In most places where the symbol X appears in the body of the procedure definition it really stands for MX.

We begin with a simple example. In it we knowingly violate the warning, given on page 183, against making assignments to formal parameters.

```
PROCEDURE M(X);
BEGIN
        X := X+7;
        RETURN X;
END;

A:=10*Y$
M(A);

        10*Y + 7
```

If we now printed out the values of A and of X we would find

Linkage questions

them unchanged (A is still 10*Y, X is clear if it was clear before). The procedure behaves as if it consisted of the lines

```
MX := 10*Y$              % automatic copying
MX := MX + 7$
RETURN MX$
```

Now consider the following tricky example:

```
PROCEDURE M(W,X);
<<      X:=123;
        WRITE X;
        WRITE W>>;

A:=10*Y$
M(A,Y)$

        123

        10*Y
```

The reader may have expected the second WRITE to produce 1230, not 10*Y, on the grounds that X and Y correspond, and that X is 123. But if we transcribe the procedure body with MX written in place of X, and MW written in place of W, we can see what happened:

```
MW:=10*Y;        % automatic copying of value of
                 % first actual parameter to MW

MX:=Y;           % automatic copying of value of
                 % second actual parameter to MX

MX:=123;         % changing MX, not Y

WRITE MX;        % writes 123

WRITE MW;        % writes 10*Y
```

If the value of Y is checked after return from the procedure it will be found unchanged (i.e., clear), because nothing in the procedure changed Y itself; what was changed was what we are referring to as MX.

The use of a local variable holding a copy of the value of the actual parameter, rather than the variable itself, is referred to as "Call by Value". It offers protection against an improperly written or improperly used procedure accidentally destroying the value of a variable in the calling environment. It also provides the mechanism that allows actual parameters to be expressions and not only variables.

4.8.1.2. The LET exception

There are times when it is intended that a procedure change the value of one of the variables that is an actual parameter. A way is provided in REDUCE to bypass the Call by Value protection. In some contexts the formal parameter X in the procedure body doesn't stand for the "copy variable" MX, but for the actual parameter's (symbolic) value itself. One of these contexts is the position before the = sign in a LET.

If X is a formal parameter of the procedure, and LET is used instead of := to make an assignment to X, e.g.

```
LET X = 123;
```

then it is the actual parameter corresponding to X that is changed. (If the actual parameter is not a clear variable, but a variable or expression equal to say P∗Q, it is as if **LET P∗Q = 123** has been executed. With P∗Q this is a meaningful LET statement; with more complicated expressions it may not be, and an error message will result.)

Here is the "tricky example" rewritten with LET:

```
PROCEDURE M(W,X);
<<      LET X=123;
        WRITE X;
        WRITE W>>;

CLEAR Y$
A:=10*Y$
M(A,Y)$

        123

        1230
```

Linkage questions 197

Let us analyze what happened this time.

```
MW:=10*Y;           % automatic copying of value of
                    % first actual parameter to MW

MX:=Y;              % automatic copying of value of
                    % second actual parameter to MX

LET Y=123;          % actually changing Y

WRITE MX;           % writes the value of MX which is Y
                    % which in turn is 123

WRITE MW;           % writes the value of 10*Y
                    % which is 1230
```

If the value of Y is checked after the procedure does its job it will be found changed to 123.

Exercise 4.8.1.

What output do you expect from the following? Guess first, then try it.

```
PROCEDURE F X $
    LET X=3 $

A := 2 $

F A;
```

As another example, we give the following procedure to store a LET rule saying that the square of the actual parameter is to be considered zero:

```
PROCEDURE NSQ X;
        LET X**2=0;

NSQ A$              % call it

A**2;               % did it work?

        0           % yes!
```

4.8.1.3. CLEAR

Consider the procedure

```
PROCEDURE CL X;
    CLEAR X;
```

This procedure is always either simply useless, or worse -- the generator of an error message! It would be profitable to see why.

After the word CLEAR just as after the word LET, the formal parameter X stands for the value of the actual parameter and not for the copy variable. Now let us examine several cases.

- CL(A), where A is clear: Then CLEAR X means CLEAR A, which accomplishes nothing. (Usually, anyway. If A is clear but is also an array name, or has weight, then these attributes are cleared.)

- CL(A), where the value of A is B: The value of A can only contain clear variables, so we know B is clear. Then CLEAR X means CLEAR B, which (usually) accomplishes nothing.

- CL(A), where the value of A is 123: We get the error message

    ***** Substitution for 123 not allowed

    because CLEAR 123 is impossible. ("Substitution" in this case means clearing.)

- CL(A), where the value of A is P∗Q: Again, we know that P and Q are clear, and also that there is no LET P∗Q = ... rule in the substitution environment (for if there were, P∗Q could not be the value of anything). So CLEAR X tries to perform (CLEAR P∗Q), i.e. delete a nonexistent rule. In this instance, nothing happens.

Now for a more useful procedure:

Linkage questions

```
PROCEDURE CNSQ X;
        CLEAR X**2;
```

Let us see if it works.

```
LET A**2 = 0$              % (might have used NSQ A$
                           %     instead)
...............
CNSQ A$                    % sometime later,
                           %     try clearing that rule

A**2;

      2
      A                    % it worked!
```

4.8.1.4. The SUB exception

Another such "non-copy" place is before the = in a SUB.

```
PROCEDURE M(W,X);
<<      W:=SUB(X=123,W);
        WRITE X;
        WRITE W>>;

A:=10*Y$
M(A,Y)$

        Y

        1230
```

Again, let's analyse this:

```
MW:=10*Y;          % automatic copying of value of
                   % first actual parameter to MW

MX:=Y;             % automatic copying of value of
                   % second actual parameter to MX

MW:=SUB(Y=123,MW); % (Note: Y here, not MX.)
                   % Changing the value of MW
                   % from 10*Y to 10*123 = 1230

WRITE MX;          % writes the value of Y which is clear

WRITE MW;          % writes the value of MW which is 1230
```

4.8.1.5. Arrays and the like

Consider the following procedure:

```
PROCEDURE AAA X;
BEGIN
        X(1) := HELLO
END;
```

Here the formal parameter is used as if it were an array or operator. What happens when AAA is called depends on the actual parameter.

Suppose the actual parameter is an array:

```
ARRAY A 10$
AAA A$
```

Then the assignment HELLO is actually made to array position A(1). Note that an assignment is made; that is, the Call by Value protection does not apply here.

```
A(1);
```

    HELLO

The same thing happens if the actual parameter is an operator:

```
OPERATOR H$
AAA H$
H(1);
```

    HELLO

If the actual parameter is neither an array nor an operator, REDUCE will ask, in the ususal way, DECLARE ... OPERATOR? and, upon affirmative response (Y) procede as above.

These examples assumed that the actual parameter was clear. Recall that an array name can also be used as an ordinary variable; we assumed that A wasn't so used.

A;

   A

If the actual parameter had a value, it is that value that determines what AAA does.

```
B := C$
ARRAY C 20$
AAA B$            % value of B is C which is an array name
C(1);
```

   HELLO

Of course if the value of the actual parameter is not a variable at all but a more complicated expression, the call on AAA will fail. It will also fail if the actual parameter is a matrix name. (The reader may try to blame this failure on the fact that AAA refers to X(1), and matrix entries must always be referred to by a double subscript; but AAA wouldn't work even if AAA contained the statement X(1,2):=HELLO.) REDUCE does not support procedures with matrix arguments, at least in the present version. Procedures with matrix arguments would have to be written on the RLISP level.

4.8.1.6. Advice

Sometimes the wisest thing to do is to run a test to determine whether in a certain context X stands for the "copy variable" or for the symbolic value of the actual parameter. For example, which is the case for the differentiating variable in DF?

```
PROCEDURE M(W,X);
<<      X:=123;
        WRITE DF(W,X)>>;

A:=10*Y$
M(A,Y)$
```

   ***** 123 invalid as kernel

Since it's the "copy variable" MX which was changed to 123 by the X:=123, we see that DF(W,X) was treated as DF(MW,MX) and not as DF(MW,Y).

4.8.2. Local variables

We have already noted, on page 190, that local variables (those listed in SCALAR declarations) can not be used as arrays, matrices, or operators. If you attempt to do so, any of a variety of error messages may be generated, including

```
FORM NIL invalid as ...

Local variable invalid as ...

NIL is not a vector

Syntax error: ...(...) invalid
```

We want to add the warning that there may be circumstances in which an attempt to use a local variable U as an array, matrix, or operator, may not be detected by REDUCE, and the procedure may appear to work but in fact it's the global variable U that is used. This has two consequences: if you had stored information in a variable with the same name, U, before calling the procedure, that information may be lost; and the procedure may work differently the second time you try it, or not work at all, because some declarations for U may be remembered from the first call.

4.8.3. The scopes of variables

Another kind of procedure linkage question is the treatment of variables global to a procedure, i.e. not formal parameters nor declared SCALAR in it. Consider the following example:

```
A := 111$

PROCEDURE ABC;
BEGIN
        SCALAR A;
        A:=222;
        . . . . . . . .
        RETURN PQR();
END;

PROCEDURE PQR;
BEGIN
        RETURN 100*A
END;

ABC();
```

What will be the answer? Which A will PQR use -- the one introduced in the first line of the example, equal to 111, or the variable A most recently introduced, equal to 222? Quite similar programming languages differ in this respect. In REDUCE the scope of a variable is determined "lexically", not "dynamically". The particular variable A that has been declared SCALAR and set to 222 inside ABC is accessible only within the lines of the text defining ABC. Since the A in question is inside the definition of PQR and not within the definition of ABC, it is taken to be the global A, the one that has been set to 111.

```
ABC();

        11100
```

4.8.4. Exit on error

This subsection deals with a very different kind of linkage question: the relationship between a user-defined procedure and the REDUCE executive system. We are referring to what happens if REDUCE, while executing a procedure, detects an error (such as attempting to divide by zero). REDUCE prints out an error message, such as

```
        ***** ZERO DENOMINATOR
```

and the procedure in which the offending operation appears is aborted. (If the procedure is the last of a chain of procedures calling other procedures, all the earlier procedures are aborted also.) The system returns to the "top level", to allow the user to enter a different command.

For example, if the command

    Y := AAA(X);

is typed in, and an illegal operation is attempted during the AAA procedure's execution, an error message (with five stars) is printed and AAA stops. Y is left unchanged. (Printouts with three stars are merely advisory, and do not indicate an aborted calculation.)

In some implementations of REDUCE the user can put error exits into his own procedures, which work just like the system-detected error exits. If he is writing a procedure ABC and has put in a test for bad data, for example, he can have the procedure stop (in the sense just described) and print out, say,

    ***** BAD DATA IN ABC

by calling on the procedure REDERR:

    IF ... THEN REDERR ("BAD DATA IN ABC");

REDERR can't itself print the actual bad data, but the programmer can include a WRITE statement before the REDERR call:

    IF ... THEN <<WRITE ..., ..., ...;
                REDERR ("BAD DATA IN ABC")>>;

The user should be aware of a danger that is a consequence of the way error conditions are treated. One often writes procedures that make some temporary change in the value of some global variable, or of the substitution environment, expecting to reverse the change when done. For example, the procedure may contain the lines

```
    .....
    LET X**5 = 0;
    .....
    CLEAR X**5;
    .....
```

If the procedure is aborted between the LET and the CLEAR, the LET rule remains in the substitution environment.

For this reason the serious REDUCE user should consider creating a procedure, perhaps called RESTORE, that contains a copy of all the CLEAR statements in all his other procedures. In case of an unexpected error, a call **RESTORE()**$ may enable him to resume calculation after correcting the error, instead of having to start over.

4.9. Procedures with GO TO

If the constructs IF ... THEN ... ELSE, WHILE, REPEAT, FOR, and so on don't suffice for describing the desired logic of a procedure, it is possible to use labels and GO TOs within the procedure body. Standards of good programming style advise avoiding this at almost any cost (but see the comment after the following example). The example, the only illustration of GO TO we shall give, presents one possible definition of a procedure to calculate factorials (there are far simpler ones!) is all we will offer. (In it we ignore the warning, given on page 183, against making assignments to formal variables.)

```
    PROCEDURE FACTORIAL N;
        BEGIN SCALAR M;
        M:=1;
    L1: IF N=0 THEN RETURN M;
        M:=M*N;
        N:=N-1;
        GO TO L1;
    END;
```

GO TO style programming can sometimes handle larger problems than programming using recursion. In one implementation of REDUCE the largest value of N that the recursive definition of FACTORIAL could handle was 31. In the same implementation the above definition easily calculated FACTORIAL 100.

While GO TO is an unconditional transfer, it can be used in conditional statements such as

```
IF X>5 THEN GO TO ABCD;
```

giving the effect of a conditional transfer.

The GO TO can only be to a label within the same procedure, and indeed only to statements on the "top level" of the procedure -- that is, not to statements inside constructions like << ... >>, IF ... THEN, WHILE ... DO, and so on.

4.10. LET rules as procedures

The LET statement offers an alternative syntax and semantics for procedure definition.

We have already seen that a procedure with no parameters,

```
PROCEDURE PQR;
   <procedure body>;
```

(where by <procedure body> we mean any single expression, or a << ... >> construction, or a BEGIN ... END construction) can be written as a LET statement quite simply:

```
LET PQR = <procedure body>;
```

To call "procedure" PQR defined in the latter form, the empty parentheses would not be used: write PQR; not PQR(); where a call on the procedure is needed.

The two notations for a procedure with no parameters can be combined. PQR can first be defined in the standard PROCEDURE form. Then a LET statement

```
LET PQR = PQR();
```

LET rules as procedures   207

would allow a user the choice of using PQR; instead of PQR(); to call the procedure.

Procedures <u>with</u> parameters can be written as LET rules also. In place of

    PROCEDURE ABC (X,Y,Z);
       <procedure body>;

one can write

    FOR ALL X,Y,Z LET ABC(X,Y,Z) = <procedure body>;

(If ABC wasn't declared to be an OPERATOR before, the system will ask if it should be. Answer Y.)

In most respects LET procedures are interchangable with PROCEDURE procedures. One difference is in the way actual and formal parameters are linked. Suppose the procedure body contains an assignment to one of the formal parameters, e.g.

    X:=123;

In the PROCEDURE case it is a variable holding a copy of the value of the actual parameter that is changed. The actual parameter is not changed, thanks to the "Call by Value" protection.

In the LET case, it is the actual parameter that is changed. (Strictly speaking, we no longer have formal parameters, but "FOR ALL" variables.) There is no Call by Value. Thus, if ABC is defined using LET, and ABC(U,V,W) is evaluated, the value of U changes to 123.

As an example illustrating the difference, we take the FACTORIAL procedure that was used to illustrate GO TO, and write it as a LET statement:

```
FOR ALL N LET FACTORIAL N =
          BEGIN SCALAR M;
          M:=1;
     L1:  IF N=0 THEN RETURN M;
          M:=M*N;
          N:=N-1;
          GO TO L1;
          END;
```

This contains the statement N:=N-1. If the user asked for the value of FACTORIAL (5), then N would <u>be</u> 5 -- not just have the value 5 -- and REDUCE would object strenuously to being asked to execute the statement 5:=5-1.

That is, in the PROCEDURE case we could merely give a warning (on page 183) against making assignments to formal parameters, saying that it usually made no trouble except to slow computation. But in the LET case, the analogous operation causes a fatal error.

Now if we introduce a new local variable, S, begin by setting it equal to N, and work with it instead of N, the LET rule works:

```
FOR ALL N LET FACTORIAL N =
          BEGIN SCALAR M, S;
          M:=1; S:=N;
     L1:  IF S=0 THEN RETURN M;
          M:=M*S;
          S:=S-1;
          GO TO L1;
          END;

FACTORIAL 10;

     3628800
```

A feature available with LET-defined procedures and not with procedures defined in the standard way is the possibility of defining partial functions.

```
FOR ALL X SUCH THAT NUMBERP X LET

                         UVW(X) = <procedure body>;
```

LET rules as procedures

Now UVW of a number (an X such that NUMBERP X is true) would be calculated as prescribed by the procedure body, while UVW of a general parameter, such as Z or P+Q would simply stay UVW(Z) or UVW(P+Q) as the case may be (assuming of course that Z, P, and Q are clear).

5. Case Studies

In this chapter we shall consider a variety of ways to solve a variety of problems. This is more than a review of what went before, because we will be pointing out unexpected complications and suggesting how to detour around them. Along the way we will also introduce a number of features of REDUCE which didn't seem to be of sufficiently general interest to warrant a place in the earlier chapters.

5.1. Find the variables

5.1.1. Find a given variable

After a series of computations and substitutions, we may want to determine if some particular variable, say XX, occurs in an expression W, or whether it has "cancelled out". We will present a number of ways to answer it, each with its own advantages and own pitfalls. There is no truly satisfactory solution!

The most obvious approach is to enter **W;** and scan the output by eye to see if XX is present. But if W is a long expression (say, many pages long) this is both time-consuming and unreliable.

A grotesque variant of this approach is to direct the "printout" to a file, and use a text editor -- outside REDUCE -- to search for XX. This method may at least seem foolproof, but it isn't. It may miss finding XX if it happens to be split between lines by a hyphenating character such as CTRL Z, as explained on page 308. It will also pick up "XX" in comments and in strings, and variables like AXXB which happen to contain the character string "XX".

Let us assume, for now, that W contains no fractions and no operator symbols -- that is, that it's a simple polynomial.

An improvement that comes to mind is to type **ORDER XX$** before entering **W;** . Then XX will be near the front of W when

Find the variables

printed, so we can spot it more quickly.

```
W := YY*(2*AA + 3*BB + 4*XX)*ZZ$

W;
```

$$YY*ZZ*(2*AA + 3*BB + 4*XX)$$

```
ORDER XX$

W;
```

$$YY*ZZ*(4*XX + 2*AA + 3*BB)$$

After the ORDER command, the XX is closer to the beginning of the printout of W. But there is still something in front of the XX, because of the ALLFAC feature. We should turn ALLFAC off. It also occurs to us, now, that there may have been previous ORDER commands, say ORDER AA,BB$, in which case the combined effect is as if we had written ORDER AA,BB,XX$, not what we want. We should first use the ORDER NIL$ command before specifying ORDER XX$ to erase the influence of any previous ORDER commands:

```
OFF ALLFAC$

ORDER NIL$

ORDER XX$

W;
```

$$4*XX*YY*ZZ + 2*YY*AA*ZZ + 3*YY*BB*ZZ$$

Success! The first variable typed out is the one we were looking for. (There is still a numeric constant, 4, in front of the XX, but this can't be helped.)

Since W is a polynomial, there is a much simpler way for us to find out whether or not XX appears in W: use the DEG function which was introduced on page 66.

```
DEG(W,XX);
```

If the answer is zero, no power of XX appears in W. If the answer is not zero, it will be a positive integer, and that power of XX is present.

Now let us remove the assumption that W contains no fractions, but retain the assumption that no operator symbols are present. Printing out W, even with OFF ALLFAC and ORDER XX, may leave XX hard to find, because (if it appears only in the denominator) it will print after the "/" symbol which may be in the middle of some line pages after the beginning of the printout. We should print the numerator and the denominator separately:

NUM W;

DEN W;

Again, the DEG function provides a simpler solution. If W is a fraction, DEG(W,XX) will only give an error message, but we can check NUM W and DEN W separately:

DEG(NUM W,XX);

DEG(DEN W,XX);

Now suppose XX doesn't appear in W directly, but only in an operator symbol such as SIN XX or M(XX,YY). If we plan to search for such an occurence by eye, ORDER XX won't do anything useful. For example, supposing H and K were operators:

S := H(AA) + H(BB) + K(2*AA + 3*BB + 5*XX);

S := H(AA) + H(BB) + K(2*AA + 3*BB + 5*XX)

ORDER XX$

S;

H(AA) + H(BB) + K(2*AA + 3*BB + 5*XX)

The ORDER command had no effect. ORDER XX$ doesn't move operator symbols on the basis of what they contain, and doesn't rearrange what's inside the parentheses of an operator symbol even if the parentheses contain XX. If we <u>knew</u> that XX appeared in, say, the context K(CC+XX), we could bring the term containing this to

Find the variables 213

the front of the expression with the command ORDER K(CC+XX)$, but without knowledge of the exact context, ORDER doesn't help. (ORDER K$ moves the variable K, if such is present, but not operator symbols of the form K(...).)

DEG is no better. If W contains K(CC+XX) then DEG(W,K(CC+XX)) will be non-zero, reflecting the highest power of K(CC+XX) which appears in W, but DEG(W,XX) will remain zero.

The Calculus provides us with a different approach. We can test to see if the derivative of W with respect to XX is non-zero. If we enter the command

    DF(W,XX);

and get output other than just a zero, we know that W contains XX somewhere.

If W is complicated, its derivative is generally more so. We can see quickly that the derivative is zero, if it is; but if it is NOT, we may have to watch the derivative printing out for a long time. Since all we want to know is whether or not it is zero, we should instead enter the command (a conditional statement)

    IF DF(W,XX)=0 THEN NO ELSE YES;

Then we get only the printout NO (there is no XX in W) or YES (there is). This of course assumes NO and YES haven't been used as variables: if they have, their values will print out, not the words NO or YES. To play safe, we should put the words in quotes:

    IF DF(W,XX)=0 THEN "NO" ELSE "YES";

Then it is guaranteed that the word NO or the word YES will print out.

Of course printing just the result of the test eliminates the printing out of the derivative, but not its calculation. This may take a long time if the original expression is long.

The derivative test is not foolproof, either. There are several sources of possible errors. First, suppose that H(XX) appears in W, but that we had earlier defined the derivative of the H operator to be zero:

```
FOR ALL X LET DF(H(X),X) = 0$
```

Then the derivative of W would be zero if XX were present only within copies of the H(XX) symbol.

Second, the value of DF(W,XX) may happen to be such that some LET rules in the substitution environment cause it to be evaluated as zero.

Third, suppose that XX doesn't in fact appear in W, but the variable YY does, and we had declared that YY depends on XX:

```
DEPEND YY,XX$
```

Then the derivative of W would not be zero. It would be some non-zero expression multiplied by the symbol (that is, kernel) DF(YY,XX).

Fourth, suppose XX appears in W, but so does YY, and the commands

```
DEPEND YY,XX$

LET DF(YY,XX) = 1$
```

had been entered. Now suppose further that W is as shown:

```
W := YY - XX$

DF(W,XX);

         0
```

We happen to get zero for the derivative, even though XX is present!

Fifth, suppose the expression contains XX**7, and we are in ON MODULAR mode with SETMODE 7. Then the derivative, which

Find the variables 215

should be 7*XX**6, will be replaced by zero.

We now arrive at what is probably the best approach, the "Change" test. Let us change XX to something else, and see if the value of W changes:

    IF SUB(XX=0,W) = W THEN "NO" ELSE "YES";

If changing XX to zero doesn't change W, then there is no XX in W to begin with. There is a danger, however, that we might create a "division by zero" situation, or something similar, in asking for SUB(XX=0,W). (For example, W might be 1/XX.) We might try SUB(XX=5,W) instead, but <u>that</u> would fail with W = 1/(XX - 5). In fact, unless we "know" W, we can't be sure to be safe no matter what number we substitute for XX.

The safest substitution is not to substitute a number at all, but a variable like XXXXXXXX which is surely a new variable and so can't have any unfortunate interactions with the rest of W:

    IF SUB(XX=XXXXXXXX,W) = W THEN "NO" ELSE "YES";

The reader may observe a mathematical relation between the derivative test and the Change test. Mathematically, the derivative is defined as a limit of quotients of the form (SUB(XX=XX+H,W) - W)/H. When we define the derivative in this way, in Calculus, we assume that the symbol H doesn't occur in W. The numerator is just a Change such as we have just discussed.

5.1.2. Finding all variables

The problem we attack here is how to direct REDUCE to list all the variables which occur in some expression W.

As in the preceding subsection, let us start with the simplifying assumption that W contains no fractions and no operator symbols.

The first step is easy: use the built-in function MAINVAR. The call **MAINVAR W** selects one variable from W, the so-called main variable. (This is the variable which is earliest in the KORDER ordering.)

    MAINVAR W;

        AA

So in this example the main variable turned out to be AA.

Glossing over a difficulty which the reader will hopefully spot without reading ahead, we proceed by deleting the main variable and determining the main variable of what remains:

    W := SUB(AA=0,W)$

    MAINVAR W;

        CC

We continue in this way until the MAINVAR is reported to be zero. This indicates that the expression contains no more variables, i.e., is now a constant, so all variables have been listed.

The difficulty is, of course, that replacing AA by zero may result in the deletion of parts of W, possibly losing some variables. The most extreme case is if W is just the product, such as AA*BB*CC*DD, of a number of variables. If we set any one of the variables to zero, W itself becomes zero. The alternative of replacing the main variable by, say, 5, fails in the same way if W should happen to be (AA-5) * (BB-5) * (CC-5) * (DD-5).

One response to this difficulty is to attempt to reduce the likelihood of this conflict by choosing "random" numbers for the values to substitute for the variables, on the theory that a number picked at random is not likely to relate badly with the given expression. We might simply pick numbers "at random", as in the example

Find the variables 217

```
SW := W$

MAINVAR NUM SW;

         AA

SW := SUB(AA=555/8,SW)$

MAINVAR NUM SW;

         CC

SW := SUB(CC=12/377,SW)$

MAINVAR NUM SW;

         DD

SW := SUB(DD=867,SW)$

MAINVAR NUM SW;

         BB

SW := SUB(BB=11,SW)$

MAINVAR NUM SW;

         0
```

Note that this time we took the precaution of working with a copy, SW, of W, rather than with W itself, to avoid destroying W. But the most significant point is that we substituted fractions, not simple whole numbers, for the variables. It seems reasonable that fractions are less likely than integers to interfere with the other variables in the expression. For example, let us relax the assumption that no operator symbols are present, and suppose that the term Z*SIN(AA*PI) is in W. Now SIN has the built-in property represented (approximately) by the LET rule

FOR ALL N SUCH THAT NUMBERP N AND DEN N = 1

    LET SIN(N*PI)=0$

If we substituted an integer, like 5, for AA, then the term Z*SIN(AA*PI) would simplify to zero, and we would lose the variable

Z (if it doesn't also appear elsewhere).

When we substitute numerical fractions for variables in W, W acquires a (numeric) denominator. MAINVAR doesn't accept a denominator, so we used MAINVAR NUM SW; instead of MAINVAR SW;. We could instead have changed the domain mode by typing ON RATIONAL$ to have W be accepted as a polynomial.

Instead of replacing the main variables, as found one by one, by numbers, we could attempt to replace them by "new" variables, the way we used XXXXXXXX in the last subsection. But we have the problem of ensuring that the new variable we insert at one stage won't be picked up as the main variable at a later stage, in place of one of the original variables. We have the KORDER command available to gain some control over what variable is chosen by MAINVAR, but unfortunately with KORDER we can only force a specific variable to come earlier in the KORDER ordering, and what we would like to do is to force the new variable to come last in that ordering, later than whatever variables were originally present. There is one hope: in many -- but not all -- implementations of REDUCE, the default ordering new variables receive places them at the end of the ordering, as long as it is early in the REDUCE run. (Later in the REDUCE run, after the underlying LISP system has gone through a so-called "garbage collection" phase, the placing of new variables becomes unpredictable.) So there is a chance that the new variables will all follow the original variables in the KORDER ordering, if all this is done early in the REDUCE session! Since this is difficult to guarantee, we will say no more about the possibility of using new variables, and return to the random numbers approach.

Instead of typing the MAINVAR and the SUB commands by hand repeatedly, we could incorporate them into a procedure.

Find the variables   219

```
PROCEDURE S W;
BEGIN   SCALAR SW, MV;
        SW := W;
        IF (MV := MAINVAR NUM SW) NEQ 0 THEN
        <<WRITE MV;
          SW := SUB(MV=RAND(),SW);
          S SW;
        >>
END;
```

This uses a procedure RAND(), which has to be supplied, for generating "random" numbers. We make up such a procedure "at random":

```
PROCEDURE RAND;
RANDSEED := (1207 + 3*DEN RANDSEED)
                     /(9812 + 5*NUM RANDSEED);

RANDSEED := 75/11$
```

(No particular statistical properties are claimed for this particular random-number generating procedure or initial seed!)

S is recursive, calling itself until MAINVAR returns zero when all the variables have been listed:

S W$

        AA

        CC

        DD

        BB

We've been speaking of MAINVAR as if it always picks out a variable. More precisely, it picks out a "kernel", which is either a variable or an operator symbol with its parentheses, like H(XX) or H(CC+XX) or, for that matter, H(). If these three operator symbol examples all appear in the expression, then they will be reported as three separate "main variables". If it isn't also used as a variable, H itself won't be reported as a main variable at any time. If XX and CC don't appear anywhere outside an operator symbol, they won't be listed either.

5.2. Dividing polynomials

5.2.1. Exact division

As an introduction to the principal topics of this section, we recall some elementary matters.

If polynomial B divides polynomial A, REDUCE will calculate A/B without difficulty.

```
A:=(X+1)*(X+2)*(X+3);

            3       2
   A  :=  X   + 6*X   + 11*X + 6

B:=(X+1)*(X+3);

            2
   B  :=  X   + 4*X + 3

C:=A/B;

   C  :=  X + 2
```

If B does not divide A, but A and B have some common factors, let us review what happens when we calculate A/B. (This was discussed earlier, on page 126.)

```
A:=X*(X+3)$

B:=X*(X+4)$

C:=A/B;

   C  :=  (X + 3)/(X + 4)
```

No problem here. But:

Dividing polynomials

```
A:=(X+5)*(X+3);
```
$$A := X^2 + 8*X + 15$$

```
B:=(X+5)*(X+4);
```
$$B := X^2 + 9*X + 20$$

```
C:=A/B;
```
$$C := (X^2 + 8*X + 15)/(X^2 + 9*X + 20)$$

The common factor (X+5) wasn't cancelled, because in order for REDUCE to discover the common factor (X+5) it must be in the ON GCD mode.

```
ON GCD;

C:=A/B;
```
$$C := (X + 3)/(X + 4)$$

```
OFF GCD;
```

5.2.2. Divisions with remainder

In algebra we often want to divide one polynomial by another getting a quotient and a remainder, the same way that in arithmetic $20/7 = 2\,{}^6/_7$, so the quotient is 2, and the remainder is 6. REDUCE always (or at any rate normally) prefers to keep expressions over a common denominator, so it would, for example, change 2 + 6/7 back to 20/7! So if A and B have no factors in common, dividing A by B does nothing interesting:

```
A:=X**5$

B:=3*X**2 + 4*X + 5$

C:=A/B;
```
$$C := X^5/(3*X^2 + 4*X + 5)$$

which looks just like what it is: A/B. We have to define our own procedure to do the job we want.

When we divide by hand in arithmetic or algebra we normally develop the quotient digit by digit or term by term, and the remainder is "what's left over". Our procedure will find the remainder first, and finding the quotient comes later!

We are trying to divide by B, which is $3*X**2 + 4*X + 5$. Finding the remainder means throwing B away as many times as possible. In algebra "as many times as possible" means until the degree is less than the degree of the divisor: that is, until there are no $X**2$ terms or terms of higher power in X.

How do we eliminate $X**2$ terms? By replacing $3*X**2$, every time we find it, by $-(4*X + 5)$. But we do not intend to insist that there be a "3" in front of the $X**2$, so we actually want to replace every $X**2$ by $-(4*X + 5)/3$.

We can use a LET statement to instruct REDUCE do this. In today's REDUCE it could be written directly using the value of B:

    LET 3*X**2 + 4*X + 5 = 0$

In older versions of REDUCE the equation had to be first solved for $X**2$ by hand:

    LET X**2 = -(4*X + 5)/3$

Either way,

    R:=A;

            R := (- 239*X - 280)/81

That's the remainder we sought!

We must tell REDUCE that we are done with the rule for replacing every multiple of $X**2$. If we forget to do this, every calculation REDUCE would make from now on would include this "simplification"! In today's REDUCE, either of the following CLEAR

Dividing polynomials 223

commands would delete the rule:

    CLEAR X**2$

or

    CLEAR 3*X**2 + 4*X + 5$

(If the reader is using an earlier or later version than REDUCE 3.2, he should check whether this statement is correct.)

Since R is the remainder, A-R is exactly divisible by B, and a division will give us the quotient Q.

    Q:=(A-R)/B;

$$Q := (27*X^3 - 36*X^2 + 3*X + 56)/81$$

Let's check the answer, by seeing if Q*B + R gives back A:

    Q*B + R;

$$X^5$$

It does!

In the above examples the polynomials contained only one variable, X, so one point which needs to be attended to in general could be conveniently neglected. Let us look at the following polynomial W, to be divided by A+B:

    W := A*B$

    LET A+B=0$

    W;

$$-B^2$$

This is "correct," because W=A*(A+B) - B². But this computation was biased, in that it was the variable A which was replaced (by -B).

Let us reverse the bias:

    CLEAR A+B$

    LET B = -A$

    W;

$$-A^2$$

This answer is equally correct. The first time REDUCE considered A∗B as a polynomial in A (namely the single term consisting of A with coefficient B); the second time, as a polynomial in B with A as the coefficient.

In general, when speaking of division with remainder we must indicate the variable we will regard as <u>the</u> variable of the polynomial.

5.2.2.1. The linear denominator case

We digress briefly to point out that if the denominator B has no X∗∗2 or higher power terms (i.e. is "linear in X") we can shortcut the LET ... ; R:= ... ; CLEAR ... ; sequence by using the SUB operation instead. For example, if B is 3∗X + 7, we can find R by writing

    R:=SUB(X=-7/3, A);

        R := (- 16807)/243

5.2.3. A polynomial division procedure

There are several awkward points encountered when trying to formalize this process as a REDUCE procedure.

Since a procedure can only return a single value, we choose to return only the remainder of the A/B division since the quotient is easily calculated as (A-R)/B when the remainder R is known. So we start:

Dividing polynomials

```
PROCEDURE POLYREM (A,B,X);
% Given A, B which are polynomials in X,
% POLYREM delivers the remainder of the
% division of A by B.
```

Notice that the variable, X, is also specified as a parameter.

Suppose B were $3*X**2 + 4*X + 5$. Suppose for now that we are working with an older version of REDUCE in which we have to separate this into $X**2$ and $-(4*X + 5)/3$. Of the tools we are already familiar with, the COEFF function is the obvious one to use for simultaneously determining the degree (DEGR=2) and the coefficient of $X**2$ (namely LC=3), but it needs an array into which to spread the coefficients. We can not declare a local variable of a procedure to be an array, so we have to use some global variable. Let us adopt the convention that the symbol CO will only be used for this and similar purposes, as a short-term storage array.

Finally, we don't know if this is the first time CO was declared, so, to avoid getting an annoying warning message "ARRAY CO REDEFINED", we begin by clearing CO (which, among other things, tells REDUCE to forget any ARRAY attribute CO may have).

(A poor alternative would be to issue the command OFF MSG$ before the ARRAY declaration. This command, which we have not mentioned before, supresses warning messages, but not error messages. If we chose to use it, an ON MSG$ would be advisable after the ARRAY declaration, in case some other warning messages are forthcoming.)

So we can continue writing the procedure:

```
BEGIN SCALAR DEGR, LC, R;
   CLEAR CO; ARRAY CO 1;
   DEGR:=COEFF(B,X,CO);
   LC:=CO DEGR;
```

Suppose B is $3*X**2 + (9/ABC)*X + 5$, which <u>is</u> a polynomial in X. This would be stored as $(3*ABC*X**2 + 9*X + 5*ABC)/ABC$, and COEFF would do nothing but print out an error message. But the denominator can be ignored, because a fraction is zero if and only if its numerator is zero, so we modify B and go on:

```
BEGIN SCALAR DEGR, LC, R;
    CLEAR C0; ARRAY C0 1;
    B:=NUM B;
    DEGR:=COEFF(B,X,C0);
    LC:=C0 DEGR;
```

(We have violated the rule that formal variables, like B, should not be modified inside procedures. We will take care of this later.)

Our next step should correspond to the LET X**2 = ... step of the manually driven process we are trying to package as a procedure. (Recall that what we are adapting is the approach that had to be used in the "older versions of REDUCE" to which we referred a few pages back. The task of writing a "modern" version of the procedure will be given as an exercise.) We have set DEGR equal to the exponent 2, so it might appear that all we need to do is to write

```
LET X**DEGR = ... .
```

Unfortunately this will store the simplification rule "If the symbol X, raised to the symbol DEGR, appears in an expression then ..."! Thus X**DEGR would be influenced by such a LET rule, but X**2 would remain as is. The reason is that what appears between the "LET" and the "=" is copied into the stored rule as is, not evaluated. (LET and CLEAR wouldn't work as intended if they cared what the present value is!)

Let us defer the solution of this difficulty for a moment and assume we have a procedure CALLLET to do what is wanted. When we directed the computation by hand, and had to type in the LET rule, the right-hand side of the "=" sign could have been calculated as X**DEGR - B/LC. So we write:

```
CALLLET(X**DEGR, X**DEGR - B/LC);
R:=A;
```

R is the result of evaluating A under the influence of the new LET rule. The next step is to clear this rule now that it has served its purpose. Attempting to do this by writing CLEAR X**DEGR will fail

Dividing polynomials

for the same reason that LET X**DEGR = ... would have failed, so for now let us assume we have a procedure CALLCLEAR to the job:

CALLCLEAR(X**2);

We will discover a flaw even in this! But to conclude:

RETURN R

Now to attack the devising of the CALLLET procedure. The solution turns out to be absurdly simple:

PROCEDURE CALLLET(U,V);
    LET U=V;

When a procedure is called, the actual parameters are evaluated, and it is their values which are passed to the procedure. So U receives the value X**2, not the symbol X**DEGR.

But wait: Why does LET U = ..., as it appears in the CALLLET procedure, not cause a rule to be stored pertaining to the symbol U, instead of its value X**2? That was the difficulty we were trying to avoid! No, there is no problem. When the symbol between the "LET" and the "=" is one of the formal parameters of the procedure, REDUCE makes an exception to its usual processing of LET, as we saw on page 196, and uses the value of the formal parameter (which equals the value of the actual parameter) and not the symbol naming that parameter! Thus CALLLET works.

One is now tempted to define CALLCLEAR analogously:

PROCEDURE CALLCLEAR (U);
    CLEAR U;

and immediately finds that it doesn't work (and sometimes produces a strange error message). What is wrong is that the actual parameter X**DEGR now has a value -- and it isn't X**2 (or whatever) but, thanks to the LET rule we stored using CALLLET, it is the "other part" of the expression B. So CLEAR (X**DEGR) doesn't work because the DEGR isn't evaluated, and CALLCLEAR(X**DEGR) doesn't work because too much is evaluated.

This dilemma can be solved by using a property of most implementations of REDUCE.

If the user inputs a command of the form LET X**5=...$, and later inputs LET X**7=...$, most implementations of REDUCE assume the former LET rule is no longer relevant, and delete it. Only one "LET X TO POWER = ..." rule for a given variable X is saved at any time. This was already discussed on page 98. Now we don't know what the power DEGR will be when the CALLLET is executed, but we know that LET X**2=0$, for example, will delete the LET rule which CALLLET stored, and replace it by one for X**2.

Now that we know that the stored rule is an X**2 rule, we can be sure that CLEAR X**2$ will remove the rule.

So instead of calling on a procedure CALLCLEAR (and trying to define a correct version of one) we can simply write

```
LET X**2 = 0;
CLEAR X**2;
```

to do the job.

One case still remains. Suppose the DEGR was 1. The LET rule which CALLLET supplies would then be LET X**1= ..., that is, LET X=... . This is not a "LET X TO POWER = " rule, but what we called a "simple LET rule", and the CLEAR X**2 will not remove it. But we can add another CLEAR to take care of this possibility (and it's harmless in the usual case):

```
CLEAR X;
```

When we inserted the line B:=NUM B; we noted that this is not proper, since in some situations changing the formal parameter can cause problems. This isn't one of those situations, as it happens, but we should obey the rules anyway. So we shall introduce a new SCALAR variable, BB, store NUM B in it and not in B, and use BB for the rest of the procedure.

We collect all the above:

Dividing polynomials

```
PROCEDURE POLYREM(A,B,X);
% Given A, B which are polynomials in X,
% POLYREM delivers the remainder of the
% division of A by B.
BEGIN SCALAR DEGR, LC, R, BB;
    CLEAR CO; ARRAY CO 1;
    BB:=NUM B;
    DEGR:=COEFF(BB,X,CO);
    LC:=CO DEGR;
        CALLLET(X**DEGR, X**DEGR - BB/LC);
        R:=A;
        LET X**2 = O;
        CLEAR X**2;
        CLEAR X;
    RETURN R
END;

PROCEDURE CALLLET (U,V);
        LET U=V;
```

Exercise 5.2.1.

Will POLYREM work if B is an expression which contains no X, for example is just a number or some variable different from X? If not, correct it.

Exercise 5.2.2.

The following alternative definition, called POLYREM2, avoids the use of the COEFF procedure. Test it, and explain how it works.

```
PROCEDURE POLYREM2(A,B,X);
BEGIN SCALAR DENOM, BB, BBB, POWR, LC, R;
    DENOM:=DEN B;
    BB:=NUM B;
    BBB:=SUB(X=1/X,BB);
    POWR:=DEN BBB;
    LC:=SUB(X=0,NUM BBB)/DENOM;
        CALLLET(POWR, POWR-BB/LC);
        R:=A;
        LET X**2 = 0;
        CLEAR X**2;
        CLEAR X;
    RETURN R
END;

PROCEDURE CALLLET (U,V);
        LET U=V;
```

For these two versions of POLYREM, we used REDUCE operations which were introduced earlier in this book in order to take B apart. The following version, POLYREM3, uses three operations which we define now:

LCOF(W,X), where W is a polynomial and X a variable, yields the coefficient of the highest power of X present in W. If X doesn't appear in W, the result is zero. LCOF stands for "leading coefficient".

LTERM(W,X) is the entire term ("leading term") of W containing the highest power of X present in it: that is, it is LCOF(W,X) multiplied by the relevant power of X.

REDUCT(W,X) is simply W - LTERM(X), the "reductum" obtained by deleting the leading term from W.

Dividing polynomials

```
PROCEDURE POLYREM3(A,B,X);
BEGIN SCALAR LC, R, BB;
    BB:=NUM B;
    LC:=LCOF(BB,X);
        CALLLET(LTERM(BB,X)/LC, -REDUCT(BB,X)/LC);
        R:=A;
        LET X**2 = 0;
        CLEAR X**2;
        CLEAR X;
    RETURN R
END;
```

Exercise 5.2.3.

Write a version of POLYREM which doesn't take B apart but essentially issues the rule LET 3*X**2 + 4*X + 5 = 0 (if that is what the value of B happens to be). Does it work if B begins with K*X**2, K a symbol rather than a number?

Exercise 5.2.4.

Analyze the following BABY procedure for doing the same thing. BABY follows the algorithm (similar to long division) taught in High School algebra:

- Divide the leading term of A by the leading term of B;

- Multiply B by the result;

- Subtract the answer from A (thus reducing the degree of A);

- Repeat this until degree of A < degree of B.

In particular, will BABY work if B is an expression which contains no X, for example is just a number? If not, correct it.

```
PROCEDURE BABY (A,B,X);
BEGIN SCALAR AA,BB,BDEG,F;
    AA := A;
    F:=1;
    BB:=NUM B;
    BDEG:=DEG(BB,X);
    WHILE DEG(AA,X) >= BDEG DO
    <<  AA:=AA - (LTERM(AA,X)/LTERM(BB,X) * BB);
        BB := BB * DEN AA;
        F := F * DEN AA;
        AA := NUM AA>>;
    RETURN AA/F;
END;
```

Exercise 5.2.5.

Compare the execution time of BABY with that of at least one of the versions of POLYREM given above.

5.2.4. The REMAINDER function

The REMAINDER function built into REDUCE can be used, under certain conditions, to accomplish what our POLYREM does. We will first show two examples for which these two functions agree.

REMAINDER(3*X**10, X**3 - 7*X + 2);

$$3*(2485*X^2 - 2752*X + 588)$$

POLYREM(3*X**10, X**3 - 7*X + 2, X);

$$3*(2485*X^2 - 2752*X + 588)$$

REMAINDER(1024*X**10,2*X**3 - 5*X + 99);

$$64*(294655*X^2 - 1990098*X + 735075)$$

POLYREM(1024*X**10,2*X**3 - 5*X + 99,X);

$$64*(294655*X^2 - 1990098*X + 735075)$$

Dividing polynomials

(REMAINDER doesn't have a third argument for specifying the variable from whose powers the polynomial is to be assumed formed. If there is more than one variable present, KORDER can be used before calling REMAINDER to specify which is to be considered the main variable.)

Here is an example in which they differ:

REMAINDER(100*X**10,2*X**3 - 5*X + 99);

$$X^6 + X^4 - X^3 + 1841585*X^2 - 12437813*X + 4591719$$

POLYREM(100*X**10,2*X**3 - 5*X + 99,X);

$$(25*(294655*X^2 - 1990098*X + 735075))/4$$

POLYREM gave an answer with a denominator. REMAINDER never does. It stopped the division process when the coefficient (now 1) of the leading term of the "remainder" (now 1*X**6) first became indivisible by the coefficient (2) of the leading term of the divisor (2*X**3). The two functions will always agree if there are no fractions anywhere to begin with, and the divisor is "monic" (that is, has leading coefficient equal to 1 or -1).

REMAINDER doesn't accept problems involving fractions:

REMAINDER(X**10/ABC, X**2 - X/5 + 7);

***** X^{10}/ABC invalid as polynomial

REMAINDER(100*X**10,(2*X**3 - 5*X + 99)/2);

***** (2*X^3 - 5*X + 99)/2 invalid as polynomial

If the mode FLOAT or the mode BIGFLOAT is set ON, and the leading term of the dividing polynomial is 1.0 or -1.0, and there are no non-numeric denominators, REMAINDER and POLYREM will give the same result.

ON FLOAT$

REMAINDER(100*X**10,(2*X**3 - 5*X + 99)/2);

$$1841593.7*X^2 - 12438112.5*X + 4594218.7$$

POLYREM(100*X**10,(2*X**3 - 5*X + 99)/2,X);

$$1841593.7*X^2 - 12438112.5*X + 4594218.7$$

Similarly, if the mode RATIONAL is set ON, and the leading term of the dividing polynomial has 1 or -1 as coefficient, and there are no non-numeric denominators, the two functions will give the same result.

ON RATIONAL$

REMAINDER(100*X**10,(2*X**3 - 5*X + 99)/2);

$$7366375/4*X^2 - 24876225/2*X + 18376875/4$$

POLYREM(100*X**10,(2*X**3 - 5*X + 99)/2,X);

$$7366375/4*X^2 - 24876225/2*X + 18376875/4$$

Analysis of why REMAINDER and POLYREM agree sometimes, and not at other times, leads to the observation that by multiplying A by a sufficiently high power of the leading coefficient of B, we can be certain to get a problem for which the two methods will always agree. This is because that high power prevents any fractions from forming (or threatening to form). The remainder computed for the modified problem is sometimes called the pseudoremainder of the original problem. We close this subsection with a version of POLYREM based on this observation. We'll call it PSEUDO.

Dividing polynomials

```
PROCEDURE PSEUDO(A,B,X);
BEGIN
        SCALAR K,AA,BB,DENA,RR;
        AA:=NUM A; DENA:=DEN A;BB:=NUM B;
        K := 1 + DEG(AA,X) - DEG(BB,X);
        K := LCOF(BB,X)**K;
        KORDER X;
        RR := REMAINDER(AA*K,BB);
        KORDER NIL;
        RETURN RR/K/DENA
END;
```

Let's analyze this. Recall that A and B may have denominators, as long as the "variable of the polynomial", X, does not appear in a denominator. We start by deleting the denominators, so as to work with pure polynomials. We compute the factor (K) by which we multiply A (now, AA). The REMAINDER function's syntax is different from that of GCD and DEG and several other functions, in that there is no provision for specifying the variable which is to be considered the variable of the polynomial. Fortunately, REMAINDER is sensitive to the KORDER-modified ordering, so by preceding the call with KORDER X we can direct that X be taken as the main variable. We are doubly fortunate in that in "KORDER X" the X, one of the formal parameters, is treated as the actual parameter and not, as is quite possible when dealing with one of the special commands like KORDER, always as the variable "X".

Before ending PSEUDO by returning as result the result of REMAINDER, appropriately corrected for the factors which were taken away or introduced, we use KORDER NIL to erase the KORDER X from the environment. Unfortunately, the user of PSEUDO must be aware that any KORDER he may have had in effect is lost.

In most of the examples tested, PSEUDO was faster -- considerably faster -- than any of the POLYREM procedures defined in the previous subsection.

5.3. LCM, GCD, and the Euclidean Algorithm

5.3.1. Least common multiple

Finding the least common multiple of two or more polynomials is easy in REDUCE. To find the LCM of, say, three polynomials A, B, C we need only add three fractions with those denominators. The denominator of the sum is the desired LCM.

There is one complication that must be avoided. If the numerators of the fractions are unluckily chosen, additional cancellation can take place in the answer. The safest course is to take, as numerators, variables which do not appear in any of the denominators: safest of all, variables which haven't been been used earlier in the REDUCE run at all.

To illustrate, let us first form two polynomials A, B with a factor X+5 in common:

```
AA:=X+5$

BB:=X+2$

CC:=X+3$

A:=AA*BB;
```

$$A := X^2 + 7*X + 10$$

```
B:=AA*CC;
```

$$B := X^2 + 8*X + 15$$

Suppose AQA and BQB are "new" variables. Then

LCM, GCD, and the Euclidean Algorithm

```
LCM:=DEN(AQA/A + BQB/B);

             3       2
   LCM := X  + 10*X  + 31*X + 30

AA*BB*CC;                    % Compare with known LCM

     3       2
   X  + 10*X  + 31*X + 30           % Correct!
```

The question arises whether some shared factors might sometimes be left in, not cancelled when the sum of the fractions is simplified, because we hadn't set the mode ON GCD. The answer is that when fractions are added, REDUCE always sets ON GCD, temporarily, at the critical stage, so we don't need to do this ourselves. If we don't want REDUCE to do this automatic ON GCD, we can change the default mode ON LCM to OFF LCM. What happens if we do this is shown below.

```
OFF LCM;

LCM:=DEN(AQA/A + BQB/B);

            4       3       2
   LCM := X  + 15*X  + 81*X  + 185*X + 150

A*B;

         4       3       2
       X  + 15*X  + 81*X  + 185*X + 150
```

Finding the LCM by adding fractions corresponds closely to the explanation of "least common multiple", in the form "least common denominator", as taught in grade school arithmetic. The following method is quite different, and doesn't depend on a wise choice of numerators:

```
ON GCD$

LCM := A * NUM(B/A);

OFF GCD$
```

If we have an array of polynomials whose LCM is desired, we

can use a generalization of this second method. Say we want the LCM of A(1), A(2), ..., A(10).

```
LCM:=1$

ON GCD$

FOR I:=1:10 DO
        LCM:=LCM * NUM(A(I)/LCM)$

OFF GCD$

LCM;
```

Still another way to obtain the LCM of two polynomials will be given at the end of the next subsection.

5.3.2. Greatest common divisor

The simplest way to obtain the greatest common divisor of two polynomials A, B is to use the built-in function GCD, by writing GCD(A,B).

There was no GCD function in some earlier versions of REDUCE. For its internal functioning REDUCE had available the operation of taking of the GCD of two polynomials, but this facility was not directly available to the REDUCE user.

There were several alternative ways to calculate the GCD of two polynomials A and B. One approach was to see what REDUCE cancels when it simplifies A/B:

```
ON GCD;

GCD:=A/NUM(A/B);

        GCD := X + 5

OFF GCD;
```

Another was to find the LCM of A and B, using one of the methods of the preceding subsection, and then calculate

LCM, GCD, and the Euclidean Algorithm

```
GCD:=A*B/LCM;

    GCD := X + 5
```

Since the GCD function is now available to us, we can use this in reverse as still another way to compute the LCM of A and B:

```
LCM := A*B/GCD(A,B);
```

5.3.3. The Euclidean Algorithm

The Euclidean Algorithm is a well-known method for finding the greatest common divisor G of two expressions A and B, considered as polynomials in a variable X. "Polynomial in X" means that X doesn't appear in the denominator, but other variables can. Since we have seen that REDUCE provides us with several simpler, faster, and more general ways to find G, we don't need to program this Algorithm if determining G is our only goal.

However, the Euclidean Algorithm also enables us to find two polynomials in X, U and V, such that G=U*A + V*B. If we are interested in determining these two polynomials, we are forced to program the Algorithm after all.

A procedure can only return a single answer, as we know. Using one of the techniques explained on page 178, in our version of procedure EUCLID we "pack" U and V together with the following specifications:

- If EUCLID is to be used, the specific variables FIRST and SECOND must not initially contain useful information, since the procedure will begin by clearing these variables.

- The answer returned by the procedure will be U*FIRST + V*SECOND.

- U can be found from the answer by setting SECOND to zero and FIRST to 1, or by using the differentiation operator DF(... , FIRST), or one of many other devices.

Then the sequence of commands

```
ANS:=EUCLID(A,B,X)$
FIRST:=A$
SECOND:=B$
ANS;
```

would produce the result of computing and simplifying $U*A + V*B$, that is, would give us the GCD.

The algorithm:

```
PROCEDURE EUCLID(A,B,X);
BEGIN
        SCALAR R,Q,AA,BB,SA,SB,SR,D;
        CLEAR FIRST,SECOND;
        AA:=A;  BB:=B;
        SA:=FIRST;  SB:=SECOND;
        WHILE (R:=POLYREM(AA,BB,X)) NEQ 0 DO
        <<ON GCD;
          Q:=(AA-R)/BB;
          OFF GCD;
          SR:=SA-Q*SB;
          D:=DEN Q;
          SA:=SB*D;  SB:=SR*D;
          AA:=BB*D;  BB:=R*D>>;
        RETURN SB
END;
```

This uses POLYREM, which was defined (in several forms) on pages 225-232. Note that EUCLID contains the command ON GCD;. Without it Q could contain factors involving X in the denominator, and the program would not work correctly. For example, if A-R turned out to be $ABC*(X+1)$, and B was $DEF*(X+1)$, then $(A-R)/B$ would be obtained in the form $(ABC*(X+1)) / (DEF*(X+1))$, with $(X+1)$ not cancelled, assuming that the default ordering or KORDER ordering at this time puts X before ABC and DEF.

With the ON GCD we can be certain all factors involving X will be cancelled from the denominator. But we don't want to do all our computations in this mode, so as soon as possible we set OFF GCD.

LCM, GCD, and the Euclidean Algorithm   241

The user of EUCLID must be aware of the fact that if for any reason he had set ON GCD before calling EUCLID the mode will be reset to OFF GCD by EUCLID. It doesn't seem to be worth the trouble to test the ON GCD/OFF GCD state as the fist step inside EUCLID, in order to restore the original state at the conclusion. If the reader disagrees, he can consider adding the test as an Exercise.

5.4. Systems of linear equations

The built-in procedure SOLVE, which was introduced on page 71, can solve systems of linear equations reasonably efficiently. In this chapter we can't hope to compete with it on the basis of efficiency. But SOLVE brings with it a large package of RLISP procedures, and, if used, cuts into the space available in the computer memory for other parts of the calculations one is asking REDUCE to do. Frequently we can do adequately with simpler tools. And if we devise our own tools, we may be able to get them to do exactly the job we wanted.

Throughout this section we shall be using the procedure CALLLET which was introduced on page 227:

```
PROCEDURE CALLLET(U,V);
    LET U = V;
```

For our first illustration, suppose we want to solve this system of equations:

$$13*X + 2*Y - Z/6 = 11,$$
$$X - 2/3*Y + 18*Z = 3,$$
$$-X + 4*Y - 12*Z = 9.$$

We need only use CALLLET to "declare" that each of these equations is true:

CALLLET(13*X + 2*Y - Z/6, 11)$

CALLLET(X - 2/3*Y + 18*Z, 3)$

CALLLET(-X + 4*Y - 12*Z, 9)$

The first CALLLET will result in a LET rule being stored for some one of the variables in the equation (we don't need to know which one), saying that it should be replaced by the appropriate expression in the other two variables. That is, the first CALLLET will take the first equation, and solve for one of the variables in terms of the others.

The second CALLLET will encounter an equation which has only two variables remaining, since one has been eliminated. It will solve for one of these in terms of the other, and store an apropriate LET rule.

The last CALLLET will solve the last equation for the remaining variable.

If we ask, now, for the values of the three variables, one of them already has its numerical value stored; in the process of answering our questions, the necessary back-substitutions to determine the numerical values of the others will take place automatically.

X;

2799/7601

Y;

23778/7601

Z;

1992/7601

If we want to use X, Y, and Z with these values in further calculations, fine; but if we want to use them as clear variables, we must not forget to CLEAR them:

CLEAR X,Y,Z;

Systems of linear equations 243

(If we had used SOLVE to solve the system, X, Y, Z would have been left clear automatically.)

Incidentally, if we <u>are</u> going to use the values which X, Y, and Z have for much further computing, it may save time if we issue the commands

X:=X$ Y:=Y$ Z:=Z$

since this will eliminate the need for repeatedly carrying out the necessary back-substitutions. They will be done once, and the <u>results</u> stored as the values of the three variables.

As a second illustration, suppose we want to fit a quadratic curve -- a parabola -- through three points (1,8), (2,5), and (7,7), and then print out the values of that quadratic for the integers 0 through 8. We type in the general equation of a parabola, three CALLLET calls, and a FOR loop to print out the values:

```
Z := A*X**2 + B*X + C$
CALLLET (SUB(X=1, Z), 8)$
CALLLET (SUB(X=2, Z), 5)$
CALLLET (SUB(X=7, Z), 7)$
FOR I := 0:8 DO WRITE I,"    ",SUB(X=I,Z)$
```

and get the answers:

         0   182/15

         1   8

         2   5

         3   47/15

... and so on, until

         8   54/5

Again, we must remember to clean up:

```
CLEAR A,B,C$
```

Our third illustration is similar to the second, but the curve to be fitted is partly defined by Calculus. Problem: find the function $f(x)$ whose second derivative is $(x-1)(x-2)^2(x-3)$ and which passes through the points $(1,0)$ and $(2,5)$.

```
A := (X-1) * (X-2)**2 * (X-3)$

Z := INT(INT(A,X),X) + C*X + D$      % C, D constants of
                                     %        integration
CALLLET (SUB(X=1, Z), 0)$

CALLLET (SUB(X=2, Z), 5)$

Z;
              6        5         4         3         2
         (2*X  - 24*X  + 115*X  - 280*X  + 360*X  +

                                             73*X - 246)/60

CLEAR C,D$
```

5.4.1. A procedure

Suppose we are given an array of linear equations for simultaneous solution. More accurately, each entry in the array is a linear expression to be made zero. Let us develop a procedure for accomplishing this.

If the array is called A, then clearly the principal step will be to execute CALLLET(A I, 0) repeatedly, with I running through the array indices.

For variety, we'll require our procedure to print out each variable and the value it should have, rather than setting each variable to that value. (Actually, each variable <u>will</u> be set to the corresponding value during the running of the procedure, but will be cleared just before the procedure terminates.)

Systems of linear equations 245

This time, we'll have to find out what variable is eliminated by each CALLLET. This we shall do by asking for MAINVAR A I -- or more exactly MAINVAR NUM A I since there could be numeric fractions in A I -- before performing the CALLLET. So we'll need an array or operator, say CO, to hold the names of the variables.

What we've produced so far is

```
KORDER NIL;
FOR I:=1:N DO
<<  CO I := MAINVAR NUM A I;
    CALLLET(A I, 0) >>;
```

The KORDER NIL is necessary because MAINVAR is sensitive to changes in the default ordering made by KORDER, but LET is not; so to make sure they agree we must not have any KORDER command in effect. (And we must warn the user of our procedure that it erases any prior KORDER.)

If we follow this by a loop

```
FOR I:=1:N DO WRITE CO I;
```

we will get the values of the N unknowns, but won't be able to tell which value goes with which variable!

Let's recapitulate. CO 1, for example, has had one of the variables, say Y, stored in it at a time when Y was clear. Subsequently, Y had a value stored in it, perhaps 7/13. If we evaluate CO 1 in the normal way, this chain of values CO(1)---Y---(7/13) is traced to the end, and WRITE CO 1 prints out 7/13.

The OFF RESUBS command was presented on page 133, but has not been used in this book so far. Its time has come! In the OFF RESUBS mode, only one step of evaluation is carried out:

```
OFF RESUBS;
WRITE CO 1;
ON RESUBS;
WRITE CO 1;
```

would first print Y, not 7/13. The second WRITE Y, with RESUBS in its normal mode, would print 7/13. So we'd get both the name of the unknown and its value.

We are hard to please, however, and would like the name of the variable to print on the same line as the value, with an equal sign between them. Each WRITE produces a separate line, so we have to find a way to combine what we have into a single WRITE. The Group statement -- here, Group expression -- is the answer. The group expression <<OFF RESUBS; CO 1>> has the value Y. (Note that there is no semicolon before the ">>", so the value of the group is the value of CO 1, computed in the current OFF RESUBS mode.)

Afterwards, the value of <<ON RESUBS; CO 1>> is 7/13.

Thus we now have

```
KORDER NIL;
FOR I:=1:N DO
<<  CO I := MAINVAR NUM A I;
    CALLLET(A I, 0) >>;
FOR I:=1:N DO WRITE
        <<OFF RESUBS; CO I>>,
        " = ",
        <<ON RESUBS; CO I>>;
```

We haven't discussed what CO should be. If CO is an array name, it turns out that OFF RESUBS doesn't consider the CO(1)---Y evaluation an evaluation step (since array-place lookup is handled by a special process), so CO(1)---Y---(7/13) is treated as a single evaluation step. We would get 7/13 printed out both times. But with CO an operator, the first printout will be Y, as desired. So we'll declare CO to be an operator.

We said we wanted our procedure to clear the variables which had been set by CALLLET before it terminates. The loop

```
FOR I:=1:N DO CLEAR CO I;
```

clears CO(1) so that CO(1) is just CO(1) and not Y; it doesn't clear Y, which is what we want. If we define a procedure

Systems of linear equations

```
PROCEDURE CCLEAR X;
        CLEAR X;
```

and attempt to use it:

```
FOR I:=1:N DO CCLEAR CO I;
```

we will get error messages to tell us that, for example, 7/13 can't be cleared. One way, CO(1) is not evaluated at all; the other way, it is evaluated too far. Again, OFF RESUBS is the answer.

```
OFF RESUBS;
FOR I:=1:N DO CCLEAR CO I;
```

We can now present the entire solution:

```
PROCEDURE CCLEAR X;
        CLEAR X;

PROCEDURE SOLVARRAY (A,N);
BEGIN
        CLEAR CO; OPERATOR CO;     %CLEAR to prevent
        KORDER NIL;                %     warning messages
        FOR I:=1:N DO
        <<  CO I := MAINVAR NUM A I;
            CALLLET(A I, 0) >>;
        FOR I:=1:N DO WRITE
                <<OFF RESUBS; CO I>>,
                " = ",
                <<ON RESUBS; CO I>>;
        OFF RESUBS;
        FOR I:=1:N DO CCLEAR CO I;
        ON RESUBS;                 % restore to normal
        WRITE "";                  % skip a line
END;
```

Now for two test runs:

```
ARRAY EEQ 3;

EEQ 1 := X + 3*Y + 5*Z - 5$

EEQ 2 := 2*X - Z - 1$

EEQ 3 := Y + Z$

SOLVARRAY(EEQ,3);

        X = 7/5

        Y = ( - 9)/5

        Z = 9/5

O                           % meaningless O since we called
                            % SOLVARRAY with ";" but it has
                            % no returned "result"

ARRAY A 2;                  % see if array name conflict
                            %   (A in procedure and here)
A 1 := X + P - 11$          %   makes trouble

A 2 := X - P - 1$

SOLVARRAY(A,2)$             % wisely used "$" this time

        X = 6               % no trouble from conflict

        P = 5
```

5.5. Series approximations to quotients

Another sense in which division is sometimes done in algebra is the one in which

```
1/(2-X) = 1/2 + X/4 + X**2/8 + X**3/16 + X**4/32 + ... .
```

This makes sense numerically if X will have a value small enough that what we left out, starting with X**5, is negligible. The right hand side, up to the three dots, is a "quotient" Q with the property that 1 - Q*(2-X) is divisible by X**5. That is, there exists a polynomial R such that

```
1 - Q*(2-X) = R*X**5.
```

Series approximations to quotients

Since (2-X) is linear, we can require R to be a constant. To make this computation in REDUCE:

```
A:=1$

B:=2-X$

R:=SUB(X=2, A/X**5)$

Q:=(A - R*X**5)/B;
```

$$Q := (X^4 + 2*X^3 + 4*X^2 + 8*X + 16)/32$$

The success of this approach depends on the fact that the denominator 2-X is linear. Unfortunately matters aren't so simple if the denominator B is of higher degree in X.

We won't attempt to explain the theory behind the method for the general case, but will illustrate it by showing how to approximate

(7*X**5)/(3*X**2 + 4*X + 5)

by a polynomial Q of degree 10 in X:

```
A:=7*X**5$

B:=3*X**2 + 4*X + 5$

C:=(A/B)/X**10$
```

So far, no surprises. Now we must write C in terms of powers of 1/X instead of X. To accomplish this we introduce a "new" variable, say Y, and set X to 1/Y:

```
CLEAR Y$

X:=1/Y$
```

We shall now find the remainder and quotient of the division problem C=AA/BB (AA, BB polynomials in Y) using one of the POLYREM procedures developed on pages 225-232:

```
AA:=NUM C$

BB:=DEN C$

R:=POLYREM(AA,BB,Y)$

Q:=(AA-R)/BB$
```

Back to the original variable X.

CLEAR X$

`Y:=1/X$`

(If you forget the CLEAR, an endless evaluation loop results: Y is $1/X$ which is $1/(1/Y)$ which is $1/(1/(1/X))$ ... .)

We must "rescale" by the correct powers of X to get the answers Q, R to fit the original A/B problem.

`R:=R*X**10*B/BB;`

$$R := (7*X^{11}*(-348*X + 3121))/15625$$

`Q:=Q*X**10;`

$$Q := (7*X^5*(116*X^5 - 1195*X^4 + 1400*X^3 + 125*X^2 - 2500*X + 3125))/15625$$

Check the answer:

`A - (Q*B + R);`

`    0`

We assemble these steps into a procedure, SERIES, with three arguments: L, corresponding to our A/B; the name of the variable (X); and the highest power desired (N, our 10). We avoid introducing the variable $Y=1/X$ by using $SUB(X=1/X,...)$ instead. We assume a polynomial division remainder routine POLYREM has been defined, or will be defined before we try using SERIES.

Series approximations to quotients

```
PROCEDURE SERIES (L,X,N);

% GETS POWER SERIES FOR L=A/B
% IN POWERS OF X UP TO AND
% INCLUDING X**N;

BEGIN
SCALAR AA,BB,C,Q,R;

        C:=SUB(X=1/X,L/X**N);
        AA:=NUM C; BB:=DEN C;
        R:=POLYREM(AA,BB,X);
        Q:=(AA-R)/BB;
        Q:=SUB(X=1/X,Q);
        Q:=Q*X**N;
        RETURN Q;
END;
```

Exercise 5.5.1.

Will this procedure work if L has no denominator or no X in the denominator? If $A=(X+1)**7$, what should be the answer to SERIES(A,X,10)? To SERIES(A,X,5)? Make any modifications to SERIES which might be necessary.

Exercise 5.5.2.

What will this procedure do if the denominator of L is X itself or is divisible by X? Is the answer reasonable?

5.5.1. Maclaurin expansion

A completely different approach to finding series approximations, not just for quotients but for more general expressions, uses Calculus: the Maclaurin Expansion (i.e., the Taylor Series in powers of X). A simple but inefficient version of SERIES, implementing this approach, follows. Note that this is the answer to part of Exercise 2.1.2 which appeared on page 44. There you were also asked to make it more efficient.

```
PROCEDURE SERIES2 (L,X,N);
FOR I:=O:N SUM
        SUB(X=O,DF(L,X,I))
          *X**I
          /FOR J:=1:I PRODUCT J;
```

Exercise 5.5.3.

Compare the times required for obtaining the series for $1/(2-X)$ using the two procedures SERIES and SERIES2. Do this for several values of N, including 5 and 10.

5.6. Families of polynomials

Problems dealing with families of polynomials are often encountered in mathematical physics and in numerical analysis. For example, we may need to express a polynomial not in terms of X and its powers, but as a combination of polynomials belonging to some specific family. We shall use the Tschebycheff polynomials to illustrate one technique, the Newton polynomials another.

5.6.1. The Tschebycheff polynomials

The Tschebycheff polynomials T_0, T_1, T_2, ... are defined by the rules

$T_0(X) = 1$,

$T_1(X) = X$,

$T_n(X) = 2 \, X \, T_{n-1}(X) - T_{n-2}(X)$ if $n > 1$.

In particular,

$T_2(X) = 2\,X^2 - 1$, $T_3(X) = 4\,X^3 - 3\,X$, $T_4(X) = 8\,X^4 - 8\,X^2 + 1$.

We can solve these equations by hand for X and its powers:

$X = T_1(X)$,

Families of polynomials

$X^2 = (T_2(X) + 1)/2,$

$X^3 = (T_3(X) + 3 X)/4 = (T_3(X) + 3 T_1(X))/4$

and so on. By combining these, we can express any polynomial in X in terms of the Tschebycheff polynomials. Our goal of the moment is to teach REDUCE how to carry out this process.

In REDUCE it is natural to write T(N) instead of $T_n(X)$. An example of the representation of a polynomial in terms of Tschebycheff polynomials in this notation is

```
8*X**3 + 6*X**2 + 5*X + 10 =
                2*T(3) + 3*T(2) + 11*T(1) + 13.
```

The symbol T must evidently be declared to be an operator. If T were an array we couldn't get answers containing the T symbols, since when an array place like T(3) is clear it is evaluated as 0, not as T(3). (While T is in some respects a reserved variable, the restrictions on its use do not include preventing it from being declared and used as an operator.)

We must replace

```
     X     by T(1),
     X**2  by (T(2) + 1)/2,
     X**3  by (T(3) + 3*X)/4,  and so on.
```

(It is not usually important to replace constants by constant multiples of T(0), and we won't do that now. However, see exercise 5.6.2.) In general, we must replace

```
     X**N  by (T(N) + U(N))/2**(N-1),
```

where U(N) is defined by

$$U(1) = 0,$$

$$U(2) = 1, \quad \text{and, for } N>2,$$

$$U(N) = 2*X*U(N-1) - U(N-2) + 2**(N-3)*X**(N-2)$$

(This is easily obtained by hand from the recursion relation defining $T(N)$, by using $T(N) = 2**(N-1)*X**N - U(N)$.)

We begin by calculating and storing an array containing the U polynomials. We shall assume we will never need to go beyond the 10th power of X, so will build the array only through the U(10) place. For convenience we will package the array construction as a procedure.

```
PROCEDURE MAKEU;
BEGIN
        CLEAR U;
        ARRAY U 10;
        U 1:= 0;  U 2 := 1;
        FOR N:=3:10 DO
            U(N) := 2*X*U(N-1) - U(N-2)
                    + 2**(N-3) * X**(N-2);
END;
```

To actually construct the array, we issue a call on this procedure:

MAKEU()$

Now that the array U is constructed, we are tempted to try to store a collection of LET rules for the powers of X:

FOR N:=1:10 DO

  LET X**N = (T(N) + U(N))/2**(N-1);

We quickly discover that this doesn't work. The rule LET X**1 = ... (that is, LET X = ...) takes precedence over the others, and we find that X**2, for example, is evaluated as T(1)**2 rather than as (T(2) + 1)/2.

(Besides this obvious error there is also a hidden flaw in our method. Each LET rule, starting with that for X**3, replaces its predecessor, so the only rules that are actually stored at the end are

Families of polynomials

those for X and for X**10.)

We revise our procedure to follow more closely the way we would do the calculation by hand. We replace the highest power of X by its expression in terms of one of the T's and lower powers of X; then the next highest; and so on until X itself is replaced by T(1). Since we are assuming no power higher than X**10 appears, we can start with N=10.

```
FOR N:=10 STEP (-1) UNTIL 1 DO
    <<LET X**N=(T N + U N)/2**(N-1);
      A:=A;
      CLEAR X**N>>;
```

This first stores the LET rule for X**10. Next, it evaluates A using this rule, eliminating any X**10 by expressing it in terms of T(10) and lower powers of X. The result is then "locked" into A. The third step deletes the LET rule since it has done its job. (One could argue that the CLEAR isn't necessary except to clear the rules for X**2 and X, since all the other rules are automatically deleted when replaced by their sucessor. But good programming style demands the CLEAR rather than depending on what may be considered a "peculiarity" of REDUCE, if we have a choice.) Now A has no powers of X beyond X**9, and the process is repeated until no X remains.

We can package this as a procedure TPEXP:

```
PROCEDURE TPEXP A;
BEGIN
        SCALAR AA;
        AA := A;
        FOR N:=10 STEP (-1) UNTIL 1 DO
            <<LET X**N = (T N + U N)/2**(N-1);
              AA := AA;
              CLEAR X**N>>;
        RETURN AA;
END;
```

Note that before TPEXP can work, the array U must have been declared and filled, by typing MAKEU()$. We don't include this inside TPEXP for two reasons. First, it is time-consuming, and repeating it, if TPEXP is called more than once, is wasteful. (REDUCE doesn't provide a program-callable test to see if a given symbol is already an array name. TPEXP would have to CLEAR U and then redeclare and recompute U.) The second difficulty is that the user would have to be reminded that TPEXP creates an array called U, to make sure he doesn't use that symbol for something else. This is less of a problem if the user is forced to define U himself as a separate step. (U can't be a local variable of TPEXP because local variables of procedures can't be declared to be arrays.)

Let us test TPEXP:

```
MAKEU() $

A := 8*X**3 + 6*X**2 + 5*X + 10$

B := TPEXP A;

    B := 2*T(3) + 3*T(2) + 11*T(1) + 13
```

There is a more efficient way to define TPEXP, using the FOR ALL ... LET ... construction.

```
PROCEDURE TPEXP A;
BEGIN
        SCALAR AA;
        AA := A;
        FOR ALL N LET X**N = (T N + U N)/2**(N-1);
            AA := AA;
        FOR ALL N CLEAR X**N;
        RETURN AA;
END;
```

We store a single rule, which REDUCE will follow over and over again until it is no longer applicable.

The author admits surprise that the FOR ALL N LET X**N = ... rule was effective on X**1, that is, on X itself; he wouldn't have expected REDUCE to consider X an instance of X**N, and thought a separate rule LET X = ... would be required. (This might indeed turn out to be the case for some other versions of REDUCE.) But this opens up another possibility: if the rule changes X to T(1), as it does, can we be sure it won't change X**5, say, to T(1)**5? We

Families of polynomials   257

have no guarantee the new rule won't be applied <u>before</u> the raising to a power. As a matter of fact, it works the way we want it to if, as is here the case, the power is a number; but if it is tried on X**K, K clear, the rule <u>is</u> applied to X first, and we get T(1)**K. So this is another instance of the unpredictability of the order of rule application by REDUCE.

The reader may have noticed that we haven't had to define the Tschebycheff polynomials in our program! What we had to define (by way of our U array) was a kind of inverse to these polynomials. To "define" the Tschebycheff polynomials we might have expected to have to express the polynomials in terms of powers of X. What we actually did was to define the powers of X in terms of the polynomials!

We need to define the Tschebycheff polynomials if we want to be able to go in the opposite direction, from a Tschebycheff expansion to a powers-of-X expression. We don't want to define values for the symbols T(N), however, because then they would not be clear and would not appear in our TPEXP answers. Instead we set the elements of an array TT to the first 10 Tschebycheff polynomials:

```
PROCEDURE MAKETT;
BEGIN
        CLEAR TT;
        ARRAY TT 10;
        TT 0:=1;   TT 1:=X;
        FOR N:=2:10 DO
            TT(N) := 2*X*TT(N-1) - TT(N-2);
END;
```

The reverse procedure UNTPEXP is then very simple:

```
PROCEDURE UNTPEXP A;
BEGIN
        SCALAR AA;
        AA := A;
        FOR ALL N LET T N = TT N;
            AA := AA;
        FOR ALL N CLEAR T N;
        RETURN AA;
END;
```

(Note that here we did the job by using a single FOR ALL rule, rather than looping through a series of N values. There was no risk in so doing, in this instance, because here there will be no subtle repetition of the application of the rule.)

Let's test it:

MAKETT()$

UNTPEXP B;

$$8*X^3 + 6*X^2 + 5*X + 10$$

which equals A.

One application of Tschebycheff polynomials is called "economizing". If a higher order polynomial is to be approximated on the interval [-1,1] by a lower order polynomial, say one of degree 4, one generally gets a better approximation by expressing the given polynomial in terms of the Tschebycheff polynomials and discarding all Tschebycheff polynomials above T(4), than by expressing it in a series of powers of X and discarding the powers above X**4.

This is easy to do in REDUCE. Perform UNTPEXP TPEXP in a substitution environment in which the higher T's are treated as zero. We give an amusing example below. Note that we will use OFF ALLFAC to make the output easier to read.

Families of polynomials

```
FOR ALL N SUCH THAT N >= 5 LET T(N)=0;

OFF ALLFAC$

FOR I:=3:7 DO WRITE I,"    ",UNTPEXP TPEXP (X**I)$
```

```
            3
   3     X

            4
   4     X

            3
   5     (20*X  -  5*X)/16

            4        2
   6     (48*X  - 18*X  +  1)/32

            3
   7     (21*X  -  7*X)/16
```

```
FOR ALL N SUCH THAT N >= 5 CLEAR T(N);

ON ALLFAC$
```

(Notice that we remembered to CLEAR the LET rule as soon as it was no longer needed.)

The reader may want to graph for himself, on the interval $[-1,1]$, the lower-order approximations to $X**5$, $X**6$, and $X**7$ which were obtained. It is the idea that such approximations are possible that the author considered amusing.

The same kind of approximate expression in terms of Tchebycheff polynomials can often be carried out to approximate a function which is not a polynomial. First get the beginning (say 8 terms) of a series expression for the function, by using the procedure SERIES defined on page 250. Then economize down to the desired degree, say four. There are two approximation steps here: the first, from truncating the power series; the second, from truncating the Tchebycheff expansion. It is not within the scope of this book to discuss the mathematical basis for making a rational decision on how many terms to take at each of the two steps; but REDUCE allows us to experiment.

Exercise 5.6.1.

Approximate $1/(3\text{-}X)$ in terms of Tchebycheff polynomials through $T(4)$. Do this several times, using as an intermediate stage series approximations going up to the 4th, 5th, 6th, 7th, and 8th powers of X. Note the effect of the intermediate approximation on the coefficients obtained.

Exercise 5.6.2.

Modify the procedures described in this section so that TPEXP 13, for example, would yield $13*T(0)$ rather than 13.

<u>Hint</u> on one approach: To find TPEXP A, first multiply A by X. Then there is no constant term. Use modified formulas for T and U in which the power of X is shifted by one: e.g., $T_0(X) = X$, not 1.

<u>Hint</u> on another approach: Calculate TPEXP A as at present. Multiply the answer by X. Apply a rule

   FOR ALL N LET X*T(N) = T(N).

Then replace X by $T(0)$.

5.6.2. The Newton polynomials

The polynomials defined by

$X^{(1)} = X$

$X^{(2)} = X * (X\text{-}1)$

$X^{(3)} = X * (X\text{-}1) * (X\text{-}2)$, etc.

are called the Newton polynomials. We will here notate them as $NW(N)$ instead of $X^{(N)}$.

Any polynomial in X can be written as a linear combination of the Newton polynomials. We want to teach REDUCE how to find the correct representation, and shall use a more direct approach than

Families of polynomials 261

we used for the Tschebycheff polynomials. (Observe that the Newton polynomials are defined individually, while the Tschebycheff polynomials were defined recursively.)

Suppose that as an experiment we try the following way to tell REDUCE how to recognize NW(3):

    LET (FOR J:=0:3 PRODUCT (X-J)) = NW(3)$

        DECLARE FOR OPERATOR ? (Y/N)

A moment's thought should reveal that we shouldn't have expected this to work, although the particular form of response could hardly have been predicted. The expression between the word LET and the = sign is never evaluated. If our LET rule had meant anything, it would have been to specify a substitution in expressions whose value contained the symbols "FOR J:=0:3 ...". REDUCE can't imagine our making such an absurd request, and instead makes a wild guess that perhaps we are trying to use the "LET operator = " form, with FOR as the OPERATOR.

Clearly we meant the value of the product to be what is considered to be NW(3). Let us copy the CALLLET procedure from page 227, and use it to define NW(1) through NW(10):

    PROCEDURE CALLLET(U,V);
        LET U=V;

    OPERATOR NW$

    FOR I:=1:10 DO
        CALLLET (FOR J:=0:(I-1) PRODUCT (X-J) , NW(I));

We first tested this using REDUCE 3.0:

```
FOR I:=1:5 DO WRITE X**I$
```

NW(1)

NW(2) + NW(1)

NW(3) + 3*NW(2) + NW(1)

NW(4) + 6*NW(3) + 7*NW(2) + NW(1)

NW(5) + 10*NW(4) + 25*NW(3) + 15*NW(2) + NW(1)

The reader can readily check that these are correct.

Before we gloat over our easy success, let us try a trivial variation on our construction (still using REDUCE 3.0): supply the LET rules in the reverse order, from index 10 to index 1.

```
FOR I:=10 STEP (-1) UNTIL 1 DO
    CALLLET (FOR J:=0:(I-1) PRODUCT (X-J)  ,  NW(I));

FOR I:=1:5 DO WRITE X**I$
```

NW(1)

$$NW(1)^2$$

$$NW(1)^3$$

and so on.

This is not what we wanted. To see why, we have to consider the mechanism REDUCE uses to build up its lists of rules.

When similar LET rules are supplied one after the other, and the similarity isn't such that the new rule causes the old rule to be deleted, the new rule is placed in <u>front</u> of the old rule. So when we supplied the rules in index order 1 to 10, the resulting rule list had, at the front, the rule for index 10, i.e. the rule involving X**10. This was followed by the rule involving X**9; and so on. When an expression was due to be checked against this list of rules, the first applicable rule was used first. For example, if the expression was just

Families of polynomials 263

X**3, the rule which said that X*(X-1)*(X-2) is NW(3) was used first, yielding NW(3) minus terms in X**2 and X. Subsequent rules eliminated the new X terms.

But when the rules were supplied with the index 1 rule last, that rule, which said X is to be replaced by NW(1), was obeyed first, replacing X and all its powers by NW(1) and its powers, leaving nothing for the remaining rules to do (because they were looking for powers of X, not of NW(1)).

When we tried the first version (index running from 1 to 10) using REDUCE 3.2, we found that it no longer worked. X evaluated to NW(1), but X**2 evaluated to NW(1)**2, X**3 evaluated to NW(1)**3, and so on. What had changed?

The answer is that in REDUCE 3.2 the rules are considered similar enough that each deletes the previous (except for the rule LET X=NW(1) which is of a different type). Of the rules for I=2 to I=10, only the last rule, for I=10, remains.

Recall that on page 82 we introduced MATCH, a variant of LET. Any number of MATCH rules, for differing powers of X, can be retained in the substitution environment.

This sounds like the answer to the difficulty, so we replace CALLLET by CALLMATCH, and try again:

```
PROCEDURE CALLMATCH (U,V);
    MATCH U=V;

FOR I:=1:10 DO
   CALLMATCH (FOR J:=0:(I-1) PRODUCT (X-J)  ,  NW(I));

X;

         NW(1)

X**2;

        2
       X
```

This is still not right. What happened is that the first rule stored was MATCH X = NW(1), and that this substitution caused the second rule stored to be not MATCH X**2 = ... but MATCH NW(1)**2 = ... Since we asked for the value of X**2 and not of NW(1)**2, the second rule was ineffective.

The trouble was that the rule for MATCH X**1 was first. If we make it be the last, we (finally) get the right answer:

```
FOR I:=10 STEP (-1) UNTIL 1 DO
    CALLMATCH (FOR J:=O:(I-1) PRODUCT (X-J)  ,  NW(I));

FOR I:=1:5 DO WRITE X**I$
```

        NW(1)

        NW(2) + NW(1)

        NW(3) + 3*NW(2) + NW(1)

        NW(4) + 6*NW(3) + 7*NW(2) + NW(1)

        NW(5) + 10*NW(4) + 25*NW(3) + 15*NW(2) + NW(1)

The moral is that the REDUCE user must always be vigilant about the possibility of unwelcome interactions between his LET rules.

Exercise 5.6.3.

Explain the last line of the following. (The LET rules were supplied in the 1-to-10 order.)

```
A := X**2;
```

    A := NW(2) + NW(1)

`A*A;`

    NW(2)^2 + 2*NW(2)*NW(1) + NW(2) + NW(1)

`A * X**2;`

    NW(3) + NW(2)^2 + NW(2)*NW(1) + 3*NW(2) + NW(1)

Families of polynomials   265

5.6.3. Orthogonal polynomials

Another way to form a family of polynomials is by requiring every pair of distinct members of the family to be orthogonal. "Orthogonal" is defined by giving the limits of integration (which we will choose to be 0 and 1) of a certain definite integral form, and a so-called "weight function" which enters into the integration. If the definite integral of the product of the weight function, the function F, and the function G is zero, F and G are called orthogonal.

Let us begin by defining DOT, a function which evaluates this definite integral:

```
PROCEDURE DOT(W,F,G);
BEGIN
SCALAR IINT, Z;
       IINT:=INT(W*F*G, X);
       Z := SUB(X=1, IINT)
          - SUB(X=0, IINT);
       RETURN Z;
END;
```

There is a well-known method, called the Gram-Schmidt algorithm, for constructing a family of polynomials P(n) such that each P(n) is of degree n, and P(m) and P(n) are orthogonal whenever m and n are not equal. We first present a REDUCE procedure GRAM for carrying out the algorithm, and explain it afterward.

```
PROCEDURE GRAM (W,NN);
BEGIN
CLEAR P,LSQ;
ARRAY P(NN),LSQ(NN);
FACTOR X;
FOR N:=0:NN DO
<<
        WRITE N,"      ",
        P N := NUM (X**N - FOR I:=0:(N-1) SUM
                DOT(W,X**N,P(I))*P(I)/LSQ(I) );
        LSQ N := DOT(W,P N,P N);
>>;
REMFAC X;
END;
```

GRAM has two parameters: the weight function W which is to

be used by DOT, and the degree NN of the highest degree polynomial P(n) which is wanted. GRAM will put the polynomials into an array P, as P(0) through P(NN).

The values of DOT computed for the squares of these functions will be needed repeatedly during the procedure, so they will be calculated only once, and then stored in another array, LSQ. (Some treatments of the subject call for rescaling the functions so that these values in LSQ would always equal to 1, but this usually leads to more complicated expressions for the functions P(n).)

P(n) starts out as X^n, and from it we subtract appropriate multiples of each of the earlier P(j)'s so as to achieve orthogonality. Since we aren't requiring our functions to be scaled in any particular way, we use NUM in order to delete their denominators and thus get simpler formulas for them.

As we construct each P(n) in turn, our procedure GRAM stores it, prints it out, and also calculates the corresponding entry for the array LSQ.

To print out the polynomials neatly, we inserted a FACTOR X statement into the procedure, to ensure that terms with like powers of X are grouped together. At the conclusion, a REMFAC X cancels its effect. (Unfortunately, if FACTOR X was already in effect when the procedure was called, that prior status is cancelled too. There is no way, in REDUCE itself, to determine what FACTOR list is in effect, so that a procedure could restore it after making a change.)

Here is an example, using the weight function $1/(X+1)$:

Families of polynomials

```
GRAM(1/(X+1), 3)$

     0     P(0) := 1

     1     P(1) := X*LOG(2) + LOG(2) - 1
                        2
     2     P(2) := 6*X *(3*LOG(2) - 2)

                  + 2*X*(4*LOG(2) - 3)

                  - 10*LOG(2) + 7
                        3             2
     3     P(3) := 30*X *(39*LOG(2) - 53*LOG(2) + 18)
                       2              2
                 + 6*X *( -9*LOG(2)  + 12*LOG(2) - 4)
                                    2
                 + 3*X*( -333*LOG(2)  + 453*LOG(2) - 154)
                              2
                 + 225*LOG(2)  - 306*LOG(2) + 104
```

For most applications we'd prefer explicit numerical coefficients, at the expense of absolute mathematical exactness, so we should perhaps carry out the calculations in BIGFLOAT mode with NUMVAL set:

```
ON BIGFLOAT$ ON NUMVAL$

GRAM(1/(X+1), 3)$

     0     P(0) := 1

     1     P(1) := X - 0.442 69504 09
                          2
     2     P(2) := 1.0*X  - 0.954 20805 86*X

                                         + 0.143 77069 6
                          3                2
     3     P(3) := 1.0*X  - 1.455 86023 7*X

                 + 0.555 86023 71*X - 0.042 64337 284
```

Let us inspect the LSQ array:

```
FOR I:=0:3 DO WRITE I,"        ",LSQ(I)$

        0       0.693 14718 06

        1       0.057 30495 911

        2       0.003 81599 512

        3       0.000 24522 093
```

The Tschebycheff polynomials can, in theory, be defined as orthogonal functions on the interval $[-1,+1]$ with weight function $1/\mathrm{SQRT}(1-X^2)$, but the REDUCE integration package can't handle the required integrals. Let us demonstrate:

```
DOT(1/SQRT(1-X**2), 1, 1);

                       2            2
        - SUB(X=1,INT(SQRT( - X  + 1)/(X  - 1),X))

                       2            2
        + SUB(X=0,INT(SQRT( - X  + 1)/(X  - 1),X))
```

INT, not being able to do the integration, returned the problem (in a slightly changed form). Then SUB, discovering this, didn't attempt to make a substitution (fortunately), but returned its problem unchanged.

(Even if the integration could be carried out for some integrals arising from using this weight function, problems could arise from "substituting" the limits because technically the integrand is improper -- infinite at the endpoints -- and a limit process, beyond REDUCE's abilities, may have to be called into use.)

5.7. Rationalizing denominators

In High School algebra we learned that fractions can always be written with no square root signs in the denominator. For example

$1/\mathrm{SQRT}(5)$ can be written as $\mathrm{SQRT}(5)/5$;

$1/(3 + \mathrm{SQRT}(5))$ can be written as $(3 - \mathrm{SQRT}(5))/4$.

Rationalizing denominators

This "simplification" isn't terribly important to REDUCE itself, since REDUCE can work with one form as easily as with the other. But it should certainly be possible to carry it out in REDUCE. Let us see how. We begin with a manually directed approach. Suppose we are given

```
A:=1/(B + 5*SQRT C)$
```

To rationalize the denominator, we must multiply both the numerator and the denominator by B - 5*SQRT C.

So we enter

```
Q:=B - 5*SQRT C$

ANS:=(Q * NUM A)/<<Q * DEN A>>;
```

Notice the superparentheses <<...>> in the denominator. This use of the grouping brackets was already illustrated on page 112, but we review it here.

If we use ordinary parentheses, REDUCE cleverly translates the formula into what amounts to

```
ANS:=((Q * NUM A)/Q) / DEN A;
```

using a strategy that often makes cancellation of common factors easier. And the strategy seems to pay off: REDUCE sees that Q divides the numerator, and carries out that division. Valuable as this may be in other problems, it is fatal here: ANS would be exactly what we started with. By enclosing the denominator in the grouping brackets <<...>> we force REDUCE to carry out the computation of the value of the group as a whole, and not separate the factors of the expression.

(We could get the same effect by calculating the denominator in a separate step:

```
BOT:=Q * DEN A$

ANS:=(Q * NUM A)/BOT;
```

The reader may well prefer this more transparent but less efficient and less elegant approach.)

Either way, we get the desired answer:

```
ANS := ( - 5*SQRT(C) + B)/(B  - 25*C)
                                2
```

In this example the denominator was simple enough, and typing in the expression for Q (the denominator with the sign of the SQRT changed) was not too difficult. But suppose the expression were

```
A:=(P+J)/(W+12*J-EE**7+L*SQRT(M-K)+99)$
```

It is far less effort (and less error prone) to have REDUCE compute the desired expression for Q:

```
Q := SUB(SQRT(M-K) = -SQRT(M-K), DEN A);

                                         7
Q := - SQRT(M - K)*L + W + 12*J - EE  + 99
```

This use of the substitution operation goes beyond its capability as described when we first introduced SUB. There we said SUB can make a replacement for a variable. Here we see SUB can also make a replacement for an operator symbol with a specific "place" filled in, i.e., a kernel.

We can now continue as before.

Exercise 5.7.1.

Rationalize the denominator of $1/(3 + SQRT(5) + SQRT(7))$. This requires two stages. The first eliminates (say) SQRT(5) from the denominator; the second deals with the result of the first step.

5.7.1. Automating the process

One aspect of automating this process is finding a SQRT in the denominator automatically. We will begin by developing a function FSQRT such that if A is SQRT 3 + SQRT 5 + SQRT 7, for

Rationalizing denominators

example, then FSQRT A will be one of 3, 5, and 7, it matters not which.

The main idea is to use a pair of LET statements:

```
FOR ALL X LET SQRT X = H X;

FOR ALL X LET H X = SQRT X;
```

where H is an operator. We will use these to change every SQRT( ) in A to H( ) and then to change them all back; meanwhile we will note what expressions occupy the parentheses, and return one of them (easiest: the last one changed) as the answer.

We will use the variable W to always hold the value of the "place" part of the latest H( ) changed, by rewriting the second of these LET rules as

```
FOR ALL X LET H X =
        <<W := X; SQRT X>>;
```

Of course these rules doesn't do anything unless we evaluate A while they are in effect:

```
FOR ALL X LET SQRT X = H X;
A := A;
FOR ALL X CLEAR SQRT X;
W := O;
FOR ALL X LET H X =
        <<W := X; SQRT X>>;
A := A;
```

Two other elements have been added. After the SQRT's have been changed to H's, the LET rule had to be cleared; otherwise both rules would remain in effect, and an endless loop of changing and changing back would result. And W was initialized to zero, so we would have a way of determining when there are no SQRT's in A.

It remains to package this as a procedure.

```
PROCEDURE FSQRT A;
BEGIN    SCALAR AA;
         FOR ALL X LET SQRT X = H X;
         AA := A;
         FOR ALL X CLEAR SQRT X;
         W := O;
         FOR ALL X LET H X =
                 <<W := X; SQRT X>>;
         AA := AA;
         FOR ALL X CLEAR H X;
         RETURN W;
END;
```

The changes made this time are minor. Since A is the formal parameter, and one is not allowed to make assigments to formal parameters, we introduced a local variable AA. To avoid trouble (an endless loop) the second time FSQRT is used, we CLEARed the rule for H( ) before exiting. And we RETURNed the answer, W.

The reader should be wondering why W was not declared a local variable.

In the chapter on Procedures we discussed contexts in which formal parameters stand for their values (the values of the corresponding actual parameters) and contexts in which formal parameters stand for themselves. Similar, but not identical, phenomena occur with local variables. In the context

```
FOR ALL X LET H X =
        <<W := X; SQRT X>>;
```

if the variable W were local the assignment to W would be treated as an assignment to the value of W. Since these lines are preceded by

```
W := O;
```

it would be treated as if were O := X, and one would get the error message

***** O IS NOT AN IDENTIFIER

We will have some comments to make on the choice of names of variables, etc., in this procedure definition, after we examine the

Rationalizing denominators

RATDEN procedure which will use it.

```
PROCEDURE RATDEN A;
BEGIN    SCALAR AA,W,Q;
         AA := A;
         WHILE (W := FSQRT DEN AA) NEQ 0 DO
         <<  Q := SUB(SQRT W=-SQRT W,DEN AA);
             AA := (Q * NUM AA)/<<Q * DEN AA>>;
         >>;
         RETURN AA;
END;
```

The WHILE condition deserves clarification. As the procedure repeatedly carries out the principal step in the rationalizing cycle, the expression (as copied into AA) will change. At each cycle, we take DEN AA and use the procedure FSQRT just defined to find some SQRT in it. If there is none, the result of FSQRT is zero, the ... NEQ 0 test will fail, and no more cycles are taken -- we RETURN the answer AA. If there was some SQRT in the denominator, we save its argument in the local variable W as an incidental step while making the NEQ test.

So when the SUB computation is ready to be made, W has an appropriate value. In this context, with W a local variable, the W in SUB(SQRT W=...) is taken to stand for its value: that is, fortunately, the SUB doesn't look for the specific letters "SQRT W" but for the SQRT of what W is equal to.

The user of the procedure RATDEN has to make sure that he also has typed in, or better, read in from a file, the procedure FSQRT. He must therefore be careful not have another procedure which happens to be called FSQRT. He must also, as it stands, make sure he has no valuable information saved in a variable called W (because FSQRT, when complete, will have set W to 0); make sure he hasn't used H for anything except possibly an operator; if an operator, make sure he has no valuable properties stored for it.

The only thing the user <u>should</u> have to be aware of is that he is using, or intends to use, RATDEN. All the other names which aren't purely internal (formal or local) should really have names with which the user is not likely to collide by accident. One convention which could be followed is to prefix the letters "RATDEN" in front of all of them:

> RATDENFSQRT instead of FSQRT;
>
> RATDENH   instead of H;
>
> RATDENW   instead of W.

Another convention is to use non-alphabetic, non-numeric characters in these names (recalling that special characters must be preceded by exclamation marks): say, FS!(!(QRT, etc.

Exercise 5.7.2.

Try RATDEN on several examples of your own choosing.

Exercise 5.7.3.

SQRT(5 + SQRT 7) has two SQRT's in it. See which one FSQRT A picks if A contains that expression. Is this the "right" answer, for RATDEN to work properly when such an A is a term in a denominator? If it isn't, what could you do about it? Hint: One way to solve the problem is to insert "IF W = O THEN" someplace.

Exercise 5.7.4.

A student proposed the following simpler version of FSQRT:

```
PROCEDURE FSQRT A;
BEGIN   SCALAR AA;
        W := O;
        FOR ALL X LET SQRT X =
                <<W := X; H X>>;
        AA := A;
        FOR ALL X CLEAR SQRT X;
        RETURN W;
END;
```

Another student simplified even this by writing a zero in place of the H X in this solution. Try these two procedures on SQRT(5 + SQRT 7) to see what's wrong with them.

5.7.2. Algebraic numbers as denominators

Up to this point we concerned ourselves with eliminating square roots from the denominator. A similar technique can be used to eliminate cube roots, et cetera. Rather than discussing this minor variation, we will examine a more general question dealing with the so-called "algebraic numbers".

We remind the reader what algebraic numbers are. If W is equal to SQRT 5, then W has the property that $W^2=5$. If REDUCE had no SQRT operator, we could introduce a variable W with all the same properties that SQRT 5 has (aside from the numerical approximation 2.23607) by supplying the rule

LET W**2 = 5$

Then W has the property that SUB(X=W, X^2-5) evaluates to zero.

If a variable has the property that the result zero is obtained when it is substituted into some polynomial, the variable is said to be algebraic, and is called a "zero" of that polynomial. (We should require that the polynomial has no proper factors, but our manipulations work even if it does.)

For example, suppose we are interested in a variable W which is to be a zero of the polynomial

A := X**3 + 5*X**2 + 37*X + 11$

just as SQRT 5 is a zero of the polynomial X**2 - 5. We'll use the procedure CALLLET defined on page 227:

CALLLET(SUB(X=W,A),0)$

Now mathematical theory tells us that A will be divisible by (X-W), the same way that the polynomial (X-3)(X-5)(X-13) -- which evaluates to zero if 5, for example, is substituted for X -- is divisible by (X-5). Let's use REDUCE to verify this:

Q := A/(X-W);

$$Q := (X^3 + 5*X^2 + 37*X + 11)/(X - W)$$

We didn't get the kind of answer expected! In fact, we got back the fraction A/(X-W).

The trouble is that division "/" is a very basic operation in REDUCE, and works on a level which is not cognizant of LET rules. When a division P/Q is to be carried out, all relevant LET rules in the substitution environment are applied to P and to Q; when the division is completed, all relevant LET rules in the substitution environment are applied to the result; but during the division process itself, no LET rules are examined. In particular, no LET rules are examined to see if the division is coming out "exact".

Fortunately, there is a related division problem which does come out exact.

Q := (A - SUB(X=W,A))/(X-W);

Since SUB(X=W,A) is zero, this is really the same division as the one with which we began. But this time, we get

$$Q := X^2 + X*W + 5*X + W^2 + 5*W + 37$$

which is the desired answer.

The reason that this works is that while the result of the SUB is zero if completely evaluated, in the version of REDUCE under which this test was run (version 3.2) the LET rule affecting W is not attended to before the division is attempted. And therefore the division succeeds, because it's a fact of algebra that for any polynomial A and any variable W, A-SUB(X=W,A) is divisible by X-W.

We referred to the fact that this was run using REDUCE 3.2, because it didn't work that way in some older versions. In these, the SUB was fully evaluated (giving 0, on account of the LET rule)

Rationalizing denominators

before the division itself was attempted, so the division failed just as if the SUB term had been absent. It was necessary to do the division (the one with the SUB term) <u>before</u> the LET rule was stored. Alternatively, one could have circumvented the LET rule for W by using, temporarily, a new variable (say WW) which has no associated LET rules:

```
Q := (A - SUB(X=WW,A))/(X-WW)$
```

The answer (which we won't bother printing) involves WW. Now substitute W for WW:

```
Q := SUB(WW=W,Q);

            2                 2
Q := X   + X*W  +  5*X  +  W   +  5*W  +  37
```

Let us not lose sight of our topic: rationalizing denominators. In our case this means rewriting a fraction which contains the variable W somewhere in the denominator in a form in which that is not so. We won't discuss the general case, in which the denominator is an arbitrary polynomial in W, but only the case of linear denominator: $U*W + V$, with no W present in U and V.

The expression Q which we found has the property that

$$A = Q * (X-W).$$

This is valid no matter what we might substitute for X. In particular,

```
  SUB(X=-V/U,A)  =  SUB(X=-V/U,Q)  *  (-V/U-W) ,

-U * SUB(X=-V/U,A)  =  SUB(X=-V/U,Q)  *  (V+U*W) .
```

Now W did not appear in A, so it doesn't appear in the left-hand side of the formula above. Therefore we can rationalize fractions with denominators of the form $U*W + V$ by multiplying the numerator and denominator by the value of SUB(X=-V/U,Q), Q having been obtained through the previously described division.

Exercise 5.7.5.

Package the above process as a REDUCE procedure.

5.8. A bug involving surds

Here we present a case study of tracking down and circumventing a "bug" in REDUCE 2.3. The problem involves fractional powers (surds).

In the section on the SOLVE function (page 71) we gave the example of finding the roots of the cubic

    A := (X-2)*(X-3)*(X-7)$

and saw that the answers given by REDUCE (which is using the so-called Cardan's Formula) had a very complicated form, instead of looking like 2, 3, and 7. One solution was

    W := (SQRT(- 3)*

                                          (2/3)
          (SQRT(- 100) + 9*SQRT(3))   -

                                          (2/3)
          (SQRT(- 100) + 9*SQRT(3))   +

                                                   (1/3)
          8*SQRT(3)*(SQRT(- 100) + 9*SQRT(3))   -

          7*SQRT(- 3) - 7) / (2*SQRT(3)*

                                          (1/3)
          (SQRT(- 100) + 9*SQRT(3))   )

Now if this equals 2, 3, or 7, then the following should evaluate to zero:

    SUB(X=W, A);

            0

A bug involving surds

It does. Or, substituting W directly into the cubic in factored form,

```
(W-2)*(W-3)*(W-7) ;
```

> 0

we again get zero.

Now suppose that for some reason we choose to compute this product of three terms in two stages. We should of course still get zero:

```
S := (W-2)*(W-3)$

S := S*(W-7);
```

> S := (6*(- SQRT(3)*SQRT(- 100)
>
> + SQRT(3)*SQRT(- 100)
>
> - 9*3 + 27))
>
> /(SQRT(3)*(SQRT(- 100) + 9*SQRT(3)))

A strange way to represent zero! This must be a "bug" in REDUCE. Our problem is to determine what happened, and to find a way to make zeros like this look like zero.

Let us concentrate on the numerator of this expression, and also discard the factor 6 which appears in front:

```
N := (NUM S)/6;
```

> N := - SQRT(3)*SQRT(- 100)
>
> + SQRT(3)*SQRT(- 100)
>
> - 9*3 + 27

Why don't the first two terms cancel? Why don't the last two, which are, in effect, 27 - 9*3?

To get a better idea of what REDUCE is actually working with, we can use the command OFF PRI$ (see page 149). This suppresses

many of the features of the printing process which make the output
"pretty", and lets us see more nearly what the internal data
representation is.

```
OFF PRI$

N;
```

```
            (1/3)   (1/6)       (1/2)              (1/2)
   ( -  3       *3          + 3       )*( - 100)

                     (1/2)   (1/3) (1/6)
              - 9*3        *3    *3         + 27
```

```
ON PRI$           % to restore the output printer to
                  %     normal mode
```

Let us concentrate on the second line. We discover that the
expression simplifier in REDUCE didn't actually multiply together the
1/2, 1/3, and 1/6 powers of 3. The output printer, with the mode
switch PRI in its normal setting, did -- hence the apparent inability of
REDUCE to determine that 9 times 3 equals 27.

Let us try to teach REDUCE to combine the exponents
properly:

```
FOR ALL X,P,Q LET X**P * X**Q = X**(P+Q)$

       ***** !=P IS NOT A NUMBER
```

REDUCE doesn't accept that LET rule. We happen to know that
internally the ** operation is called EXPT, so we try using that:

```
FOR ALL X,P,Q LET EXPT(X,P) * EXPT(X,Q) = EXPT(X,P+Q)$

       ***** !=P IS NOT A NUMBER
```

No improvement!

Further experimentation shows that if REDUCE is asked to
simplify X**P * X**Q, and P and Q are fractions with "unrelated"
denominators (neither one divides the other), then the expression
simplifier doesn't combine them.

A bug involving surds 281

```
5**(1/5) * 5**(1/7) - 5**(12/35);

       (12/35)     (12/35)
   - 5         + 5

5**(1/5 + 1/35) * 5**(1/7 - 1/35) - 5**(12/35);

       0
```

Maybe we can just tell REDUCE what to do in this particular case.

```
LET 3**(1/3) = (3**(1/6)) ** 2 $

S;

       0                    % success!
CLEAR 3**(1/3) $            % remove the rule
```

But it's not very practical to enter a rule for each specific problem which arises.

Let's try something else. We know that the normal (ON PRI$) printing process <u>does</u> combine exponents properly. Maybe <u>any</u> procedure does. Let us define an "identity" procedure ID, which simply returns whatever it's given, and pass S to it:

```
PROCEDURE ID X; X;

ID S;

       0
```

It works.

So we have discovered that, when using REDUCE 3.2, if there is reason to suspect that an expression, say Z, containing surds (fractional powers) may not be simplified as much as possible, we should use ID(Z) instead (assuming, of course, that we have included the definition of this procedure ID in our run).

Of course <u>this</u> bug will probably be corrected in later versions of REDUCE. But other bugs may always turn up, and the user of

REDUCE (or any other computer program) should be aware of this possibility, and be ready to investigate -- as above -- unexpected or unreasonable answers. It is reasonably safe to expect that most remaining errors in REDUCE are of the kind that some possible simplifications have been overlooked. Answers are not likely to be actually in error, with the exception of the single type of error resulting from the assumption REDUCE makes that, for example, (A**6)**(1/2) is A**3. (It may be -A**3.) When the author encountered the "bug" which is the subject of this case study, he first assumed that the trouble came from some form of this mathematical ambiguity!

5.9. Noncommuting symbols

In the section on matrices it was pointed out that REDUCE knows that if A and B are matrices, then A*B and B*A are not in general equal, so that B*A is not to be automatically replaced by A*B (or vice versa). That is, multiplication of matrices is not commutative.

There are other occasions in mathematics when we don't want to treat the multiplication sign "*" as commutative. For example, in the algebraic system known as the quaternions, there are three special symbols I, J, K with properties such as I*J = K but J*I = -K.

In REDUCE we can declare that certain symbols are not to be interchanged in multiplication. Unfortunately, this applies only to operator forms. We can't say that I and J shouldn't commute, but we can say that I() and J() shouldn't commute:

```
OPERATOR I,J$

NONCOM I,J$

U := I()*J()$

V := J()*I()$

U-V;

      I()*J() - J()*I()
```

(We couldn't have made an equally direct demonstration showing that matrices don't commute, with a final line A*B - B*A, because matrix names never appear in answers, only their entries do: that is, matrices can't be clear.)

More generally, if there is a declaration NONCOM P,Q,R,S$, or a series of declarations NONCOM P$, etc., then whenever a term appears in which say P(123) and S() or S(456) or S(AA,BB,CC) are multiplied together, the order of those particular symbols is not disturbed (while other factors in the same terms may be rearranged freely by REDUCE).

5.9.1. Quaternions

As we just saw, NONCOM can be used to declare that the usual law of algebra, that the factors of a product can be interchanged, is to be suspended for certain operators; then LET rules can be given to specify the way "*" should behave for those operators. We've already mentioned the system of quaternions. To list, explicitly, the multiplication rules for the three basic quaternions I,J,K would require a collection of nine LET rules. We can state the rules more compactly if we use a single operator symbol, say Q, with different "place" entries filled in, to represent I,J,K: namely Q(1) through Q(3).

```
OPERATOR Q$

NONCOM Q$

FOR ALL I LET Q(I)**2 = -1$

FOR ALL I,J SUCH THAT J = I+1 LET Q(I)*Q(J) = Q(6-I-J)$

FOR ALL I,J SUCH THAT I>J LET Q(I)*Q(J) = -Q(J)*Q(I)$

LET Q(1)*Q(3) = - Q(2)$
```

These LET rules express that I^2, J^2, and K^2 are all equal to -1; that I*J = K and J*K = I; that interchanging distinct basic quaternions in a product changes the sign; and finally, not readily falling into the pattern of the other rules, the rule I*K = -J. Now to check:

```
FOR I := 1:3 DO
   FOR J := 1:3 DO
      WRITE I," ",J,"     ",Q(I)*Q(J)$

1   1      (-1)

1   2      Q(3)

1   3    - Q(2)

2   1    - Q(3)

2   2      (-1)

2   3      Q(1)

3   1      Q(2)

3   2    - Q(1)

3   3      (-1)
```

We may be reluctant to type in Q(1) when we mean I, and so on. Can we simplify input?

```
II := Q(1)$

J := Q(2)$

K := Q(3)$

S := 5*II*K + 7*J*II + 66*II*J + 66*J*II;

    S :=  - 7*Q(3) - 5*Q(2)
```

The value printed for S demonstrates that REDUCE replaces II, J, and K by their "values" Q(1) through Q(3) before it considers putting the symbols II, J, K in a standard order in the product, so this input trick works. (Of course we couldn't use a plain "I" because it's a reserved name, so had to substitute "II".)

Now let us transform the <u>answer</u> into the II, J, K notation:

```
CLEAR II,J,K$

Q(1) := II$

Q(2) := J$

Q(3) := K$

S;
```

                    - 5*J - 7*K

Presumably if we are working with quaternions we'll have to do these two translation tricks repeatedly, so we should package them for convenient use.

```
LET INIJK =
<<      CLEAR Q(1),Q(2),Q(3)$
        II := Q(1)$
        J  := Q(2)$
        K  := Q(3)$
0 >>$

LET OUTIJK =
<<      CLEAR II,J,K$
        Q(1) := II$
        Q(2) := J$
        Q(3) := K$
0 >>$
```

Since no parameters are involved, we defined the packages as ordinary variables, using LET, instead of procedures (so as to simplify typing: use INITJK$ instead of INITJK()$). A zero has been placed before the closing ">>" symbols to to make explicit the fact that the "values" of INIJK and OUTIJK are unimportant.

```
INIJK$

W := II*(13*II + 17*J + 19*K)$

OUTIJK$

W;
```

                    - 19*J + 17*K - 13

We can even combine this into a single line:

```
<<INIJK>> + <<II*(13*II + 17*J + 19*K)>> + <<OUTIJK>>;
    - 19*J + 17*K - 13
```

The superparentheses assure that REDUCE will calculate the three parts in the order given. The "values" computed for the first and third parts are zero, so they don't contribute anything to the sum; but in the process of "computing" their zero value the desired side-effects take place in the right order. (We aren't really recommending this shortcut.)

5.9.2. Steenrod Squares

In the branch of mathematics called Algebraic Topology, certain noncommuting symbols called the Steenrod Squares are sometimes dealt with. We don't expect many readers to be familiar with these symbols, but the example may be instructive nevertheless.

The so-called Adem Relations, which enable some products of Steenrod Squares to be simplified, involve binomial coefficients, so we need a procedure for computing them.

```
PROCEDURE BC(M,N)$
    IF M<O OR N<O OR M<N THEN O
        ELSE FACT M/(FACT N * FACT (M-N))$

PROCEDURE FACT N$
    FOR I:=1:N PRODUCT I$
```

Several comments are in order. The Binomial Coefficient function is a function of two arguments, say BC(M,N). It is normally assumed that M and N are non-negative, and that M is greater than (or equal to) N. The Adem Relations sometimes ask for BC values which violate these conditions; in these cases BC is required to return the value zero. This is assured by the "THEN" case of the IF ... THEN ... ELSE in our definition.

There are many reasons why the Binomial Coefficient, in the normal case, should not be calculated by using the formula we gave which involves the factorial of three numbers. The formula for BC(100,1) involves the huge numbers FACT(100) and FACT(99), but cancellation takes place and the final answer is only 100. This answer could have been found by a more economical method. In programming languages like FORTRAN, BASIC, or Pascal, factorials rapidly reach values which exceed the arithmetic capacity of the computer; even when not too large to be represented, roundoff error affects the accuracy of the answer. And calculating factorials of large numbers is slow.

The sizes of the factorials is no problem for REDUCE, since it works with unlimited precision integers; and there is no roundoff in integer computations. The concern about speed is valid, until we consider that REDUCE is not intended as a high-speed computing engine but as a tool of convenience. We will not be asking for thousands of binomial coefficients! The convenience of defining the binomial coefficient function by a simple formula outweighs the inefficiency of the calculation itself.

The conventional notation for the Steenrod Squares is Sq^n, but naturally we will use SQ(N) in REDUCE. So SQ must be declared to be a non-commuting operator symbol, with the Adem Relations to simplify certain products of them:

```
OPERATOR SQ$

NONCOM SQ$

FOR ALL A,B SUCH THAT A < (2*B) LET SQ(A)*SQ(B)=
    FOR C:=0:A SUM
        IF REMAINDER(BC(B-C-1, A-2*C),2) = 1
            THEN SQ(A+B-C)*SQ(C)
            ELSE 0$
```

All that the Adem Relations ask about the binomial coefficients is whether they are even or odd. The REMAINDER(,2) = 1 test determines that. The rule expresses the product of two SQ()'s satisfying the SUCH THAT condition as a sum of zero or more "simpler" products of SQ()'s. The rule is automatically applied to all the terms of the answer, recursively, until none of the terms satisfy the SUCH THAT condition.

We illustrate with a few examples:

```
SQ(4)*SQ(10) ;
        SQ(12)*SQ(2)
SQ(3)*SQ(26) ;
        0
SQ(5)*SQ(14) ;
        SQ(19)*SQ(0)  +  SQ(17)*SQ(2)
```

Exercise 5.9.1.

The Adem Relations should apply to products in which A=B (if A is greater than 0). Does our LET rule version do so? If not, add a second LET rule to take care of this case.

6. Running REDUCE

In this chapter we present the unfortunately necessary details of handling the communication between you who wants to use REDUCE, and the computer in which REDUCE resides. Included are such mundane topics as how to use files created by a text editor (or by another REDUCE run) to cut down on typing and retyping; correcting (editing) input lines, or procedure definitions, or files; setting the output line length to meet the requirements of display screen or printer; turning procedures into machine language form (compiling) for faster execution; and so on.

The material in this chapter is highly sensitive to differences in implementation (especially to differences between computers and between operating systems). What is written here emphasizes the implementations for the IBM-PC by Northwest Computer Algorithms, and the implementation for the DEC 20 with the TOPS 20 operating system that was widely distributed in the past, but its usefulness is not limited to these. We hope the reader will be able to adapt the material to the environment in which he is working. Some of the features we describe may be absent; some may operate differently; and other features, not described here, may exist. If a REDUCE User's Guide is available for your system, you should consult it while you are reading this chapter.

6.1. The basics

The first step in using REDUCE is to turn on the computer or terminal, log in if required, and call up REDUCE. How this is done varies from system to system. On some, you merely type **REDUCE** and strike the RETURN key; on others, you may have to type **CD REDUCE** or some other incantation first, and strike RETURN, before typing **REDUCE** and striking RETURN again. The possibilities are endless. If you don't know, ask.

When REDUCE has been conjured up successfully, after a pause an acknowledgement line will print out informing you which version of it you have. Then, possibly after another pause, the prompt symbol

"1:" will appear. (Older versions of REDUCE didn't give numbered prompts, but only a symbol such as "*".)

In response to each prompt, you can enter

- a statement or command, such as an assignment statement or a mode-setting ON or OFF command. End it by typing a dollar sign and striking RETURN. (If you want to see the result, if any, of the statement or command, type a semicolon instead of a dollar sign before striking RETURN.)

- or an expression whose value you want to see. End it with a semicolon, and strike RETURN.

- or the definition of a procedure. End it by typing a dollar sign and striking RETURN. (If you type a semicolon instead of the dollar sign, the name of the procedure just defined will print out.)

If the item is completed in one line, the next line will have the prompt "2:", and so on. If the item doesn't fit conveniently on one line, you can start a new line by striking RETURN at any reasonable place: but not in the middle of the name of a variable, not in the middle of the sequence of digits representing a number, and not in the middle of text in quotes (" "). The number in the prompt doesn't change until the item is completed by entering the final semicolon (or dollar sign) and RETURN.

If you used semicolon and RETURN, the result will print out before the next prompt symbol appears.

If you assign a variable a value using a ":=" assignment statement, or a LET, and use that variable in a subsequent expression, that value will be used in evaluating the expression. If later in the session you assign the variable a new value, then from that point on the new value will be used.

The basics

The same is true for procedure definitions. Procedures can be defined and redefined. Any time it is necessary to carry out a procedure while evaluating an expression, the most recent definition will be the one used.

Either QUIT$ or BYE$ can be typed to end the REDUCE session. Of course if your computer system has a standard way to abort programs (such as the CTRL C which works on many systems), that can be used instead.

The reader is invited to replicate the following REDUCE session at his own computer or terminal. If you make a typing mistake, the correction tools usually available on your system will probably work: for example, on some systems the last few characters typed can be "erased" by hitting BACK SPACE enough times, on other systems by hitting DELETE. In the IBM-PC version, CTRL C will tell REDUCE to ignore what has been typed so far on the current line. In addition, on some systems REDUCE allows CTRL G to be used to "erase" the entire statement or command or definition that has been partially typed, even if you are already on the second or a later line of the item.

```
@REDUCE                          Call in REDUCE.
(REDUCE version xxxxx)              (Acknowledgement)
1: X:=(Y+Z)**2;                  Set X to (Y+Z)².
         2              2
X  :=  Y   + 2*Y*Z  +  Z         The expanded result
                                    is printed.

2: DF(X,Z,2);                    Find d²X/dZ².

2                                Here's the result.

3: PROCEDURE FACT N;             Now define a
3:      BEGIN SCALAR M, S;          procedure for calculating
3:            M:=1; S:=N;           factorials.
3:       A:   IF S=0 THEN RETURN M;
3:            M:=M*S;
3:            S:=S-1;
3:            GO TO A;
3:      END$                     ($, so name "FACT" is
                                    not echoed)
```

4: 2**FACT 3;	Try it in an expression.
64	The answer is $2^6 = 64$.
5: FACT 120;	This is a big number:

66895029134491270575881180540903725867527463331380298102956713523016335572449629893668741652719849813081576378932140905525344085894081218598984811143896500059649605212569600000000000000000000000000000

6: QUIT$

6.2. IN from files

If some or all of the commands, definitions, etc. to be entered are at all complicated, it is advisable not to try to type them directly, but to store them in a file (using some text editor) <u>before</u> starting the REDUCE run.

On many systems, file names have two parts: the name proper, and a file type or "extension", the two separated by a period. On such systems, it's conventional to designate files intended for reading by REDUCE with the extension RED.

For example, for the illustration at the beginning of this chapter we could have created a file ABCD.RED before calling in the REDUCE system. If this were done on a system with a text editor named CREATE that numbers the lines being input, this might look like this:

IN from files

```
@CREATE ABCD.RED
00100    X:=(Y+Z)**2;
00200    DF(X,Z,2);
00300    PROCEDURE FACT N;
00400            BEGIN SCALAR M, S;
00500            M:=1; S:=N;
00600      A:    IF S=0 THEN RETURN M;
00700            M:=M*S;
00800            S:=S-1;
00900            GO TO A;
01000            END;
01100    2**FACT 3;
01200    FACT 120;
01300    QUIT$
01400    END$
```

At this point one would do whatever is necessary to leave the CREATE editor and save the file under the name ABCD.RED.

To use this file, enter REDUCE and issue an IN command. On most systems the file name (with its extension) has to be enclosed in double-quote marks, as shown below. On some systems, if the file name does not have an extension, the file name can be typed after IN without quotation marks.

```
@REDUCE
(REDUCE version xxxxx)

1: IN "ABCD.RED";
```

Now the lines making up the file ABCD.RED will appear on the screen just as if they were being typed directly, except that no prompts are shown. Each command, etc., is obeyed as it comes. Since this particular file was typed with a semicolon after each command, the "value" of each prints out below the display of the command itself.

When the end of the file is reached, the next prompt (in this case, "2:") appears. On some systems an END$ is required as the last line of the file, to assure proper handling of subsequent input; more on this shortly.

If we don't want to have the result of a command or value of an expression printed automatically, the file should have a dollar sign $ terminator after the command or expression, instead of a semicolon.

If we end the IN command with a dollar sign (and RETURN) instead of a semicolon (and RETURN), the contents of the file itself would <u>not</u> appear on the screen, but the results (the value of each semicolon-ended command in the file) would, one after another.

If the command **OFF ECHO$** appears in the file, the rest of the file from that point on will not be shown even if a semicolon had followed the IN command. Conversely, **ON ECHO$** will ensure listing from the point where it appears in the file even if $ was used. Sections of the file can thus be selectively displayed by the inclusion of **ON ECHO$** and **OFF ECHO$** commands.

The IN command can be used at any time during a REDUCE session, not just at the beginning. For example, if a certain series of commands is to be performed several times during a session, instead of incorporating them in a formal procedure one might place them in a file and "IN" the file whenever needed.

A single IN command can be used for reading several files one after the other:

    IN "ABCD.RED", "UVWXY.RED";

Furthermore, a file to be read using IN could itself contain IN commands to read other files.

Recall that if one types in a statement such as **A(5) := 123;** directly, and A has not been declared an array or an operator, REDUCE asks "DECLARE A OPERATOR?" and waits for a response. When input is from a file, through "IN", REDUCE doesn't ask such a question, but <u>assumes</u> an affirmative answer and simply reports "A declared operator".

<u>Caution</u>: If an array declaration like **ARRAY A 20$** appears in a file to be read repeatedly, it should be preceded by a **CLEAR A$** to

IN from files   295

prevent getting an "ARRAY A REDEFINED" message each time after the first. Alternatively, the array declaration should be removed from the file and simply entered by hand before the first reading of the file.

We said the file is processed until the end of the file is reached. However, it terminates sooner if an END$ not matched to a BEGIN is encountered. Therefore it doesn't hurt to put an END$ command at the end of the file (and indeed, as we already noted, in some implementations it is required). Even in implementations in which the final END$ is not required, including it is a useful safety device. For suppose the file contains a number of BEGIN ... END pairs (presumably in procedure definitions), but one of the END's is omitted by error. An extra END$ at the end of the file closes the bracket, and prevents possibly very puzzling behavior when manual typein is resumed. This puzzling behavior takes such varied forms as messages "END-COMMENT NO LONGER SUPPORTED", the prompt number not advancing, or one of the procedures defined in the file always giving as result the symbol EOF instead of what its definition would seem to call for.

The command END$ can also be placed in the middle of a file, temporarily, to "cut off" the subsequent portion of the file in order to facilitate debugging the earlier portion.

6.2.1. REDUCE.INI

On some systems REDUCE makes a test, at the beginning of every session, to see if there is a file named REDUCE.INI in the user's directory. If there is, it automatically initiates an IN "REDUCE.INI"$ action. Typically, one would keep a collection of one's own procedure definitions in this file, which would be automatically loaded whenever a REDUCE session is started, thus enabling one to have a customized environment to work in. One should take care to have all definitions (and other commands) in this file terminate with $ signs, not semicolons, for otherwise the names of the defined procedures, and results of other commands, will print out placed confusingly between the version identification printout and the first prompt line.

6.2.2. PAUSE, CONT, and DEMO

Normally, once an IN is initiated, the entire file of commands (by which word we mean to include procedure definitions and expression evaluations as well) is read in and processed. If one wishes to read in one group of commands, perform some computations by direct manual typein, then read in some additional commands, one could put the two lists of commands on different files and first "IN" one, and later "IN" the other.

An alternative is to put the command PAUSE$ in the file at the end of the first list of commands. When PAUSE$ is executed, processing of the file is interrupted, and the question Cont? (Y or N) appears on the screen. A "Y" response -- followed by RETURN -- continues processing of the file immediately. An "N" response -- followed by RETURN -- allows one to input other commands, etc., manually, or even to "IN" a different file. Then, to resume processing the file that was interrupted, enter a CONT$ command manually. (The CONT$ command can only be used if some file has been interrupted by a PAUSE$ within it. If used at other times, a "NO FILES OPEN" error message is generated.)

If you are demonstrating REDUCE to someone, and have the demonstration command sequence on a file, you may want REDUCE to pause after each command's execution so that you can explain what's happening. It is not necessary to put a PAUSE$ command after each command: simply enter the mode-setting command ON DEMO$ before entering the "IN" command (or begin the file itself with ON DEMO$). Then REDUCE automatically stops after printing each result. Just strike RETURN to continue to the next command on the file. The ON DEMO$ command has no effect on the way REDUCE handles manually typed input. But if you want the next file read in to be processed without interruption, remember to issue an OFF DEMO$ command.

IN from files

6.3. Making corrections

What is to be discussed in this section is probably the most implementation-dependent aspect of REDUCE. Correcting typing and editing files requires considerable cooperation between REDUCE and the operating system of the computer being used, and it has not been possible to implement all desirable correcting modes on all computers supporting REDUCE. As we said before, if a User's Guide for the implementation of REDUCE on your computer is available, consult it for details. If not, then experiment!

6.3.1. Correcting as you type

This is mostly a review of some remarks made on page 291:

- Hitting BACK SPACE or perhaps DELETE will probably cause the computer to ignore the last character on the line you are currently typing. Two BACK SPACEs or DELETEs will delete the last two characters, and so on.

- To delete the entire line so far typed, either hit BACK SPACE or DELETE enough times, or see if CTRL X or CTRL U or CTRL C does the job on your system.

- To delete not only the entire current line, but the entire current statement (which may be more, if the statement takes more than one line), try CTRL G. On most keyboards this will also sound a beep -- the "bell" signal.

6.3.2. Correcting the previous statement

Suppose you (or REDUCE) discover an error in a statement after you have already terminated the statement (with a $ or ; followed by RETURN). If REDUCE didn't catch the mistake, then your erroneous statement was executed, so you'll have to figure out what harm, if any, was caused, and how to repair the damage. But aside from that, you now want to have the correct statement processed.

If the statement is short and simple enough, the easiest thing to do is to retype it by hand. But in some REDUCE implementations you are given the option of editing what you typed before.

How to accomplishing this editing, if possible at all, differs from implementation to implementation. To give the reader a taste of what's possible, we describe the editor that is provided in the IBM-PC and DEC 20 implementations.

Type the command ED$. In the IBM-PC version you may get the message "ED not supported". If so, enter LOAD "CEDIT";, wait until the disk stops spinning and you get a new prompt, and try again. (LOAD is explained later, on page 311.) REDUCE will then reprint the statement for you, and activate a simple so-called string editor for correcting the error. After editing is complete, typing E (for Exit) causes the corrected line to be reprocessed as if it had just been typed.

On the IBM-PC this editor makes use of CTRL D. The DEC 20 version uses the key marked ESCAPE or ESC, available on most terminals.

While using this string editor it is necessary to visualize an imaginary cursor, or pointer, which is initially positioned just to the left of the first character of the statement. It has to be made to sweep across the statement as corrections are made. The most commonly used editing commands are as follows:

- To check on the location of the invisible cursor, use P, which prints the part of the statement between the cursor and the end (but does not itself move the cursor). If you discover that the cursor is further along the statement than you expected, there is no way to back up partially: you must use the B command to move the cursor all the way back to the beginning. The corrections you have made remain. (We did say this was a simple string editor!)

- To move the cursor one place to the right, hit the space

Making corrections 299

bar. (If the cursor had been at the beginning of the line, the new position would then be just to the left of the second character.)

- F, followed by a character, positions the cursor just to the left of the next occurrence of that character in the statement. F doesn't look at the character that is initially immediately after the cursor; it starts its search after that character. If the desired character can't be found, the cursor remains where it was.

- I, followed by a string of one or more characters and then CTRL D [or ESCAPE], (don't forget the CTRL D or ESCAPE!) inserts those characters between the cursor and the next character. The cursor doesn't move. Caution: This means that if you have just inserted the character string ABCDE, and then use FC to find the "next" letter C, F will actually find the C that was just inserted.

- D deletes the single character following the cursor. (If used immediately after an insertion using I, the first character just inserted would be the one deleted!)

These commands can be typed one per line, with a RETURN after each, or collected several to a line. In the latter case they are not obeyed until RETURN is struck, so the BACK SPACE or DELETE key can be used to make corrections in the sequence of commands until then.

For example, supposed you discover you had entered

CHECKS := MO + CAT + CDOG;

instead of

CHECKS := MOUSE + CAT + DOG;

Entering the command ED$ results in a ">" prompt symbol. Now we first want to Find the first "O", Space once to get to the far side of that "O", and Insert "USE" (to form the word "MOUSE"). When we hit CTRL D or ESCAPE after typing the "USE", a new ">" prompt is printed. We continue: Find the first "+" after this point, Find the next "+", Find the next "C", and Delete it. Now to check that these corrections were made correctly, go Back to the beginning and Print. Finally, enter an E. This is what the entire process looks like:

```
51: CHECKS := M0 + CAT + CDOG;

52: ED$

CHECKS := M0 + CAT + CDOG;        % Statement to be edited

>F0 IUSE^D>F+F+FCDBP              % Editing commands,
                                  % ending with B, P, and
                                  % RETURN for
                                  % proofreading

CHECKS := MOUSE + CAT + DOG;      % Looks OK, so ...
>E                                % ... enter E (for Exit)
                                  %   and strike RETURN

CHECKS := MOUSE + CAT + DOG;      % Corrected statement is
                                  % automatically re-input
                                  % into REDUCE
```

The Q command is rather useful. If you have gotten too confused to continue, Q (followed by RETURN) exits from the editor without saving or reprocessing the partially corrected (or miscorrected) statement. You can the issue ED$ again, to start correcting the original form of the statement over again, or take the other alternative and type the desired statement by hand.

The complete list of commands follows:

```
B                  Back     move pointer to beginning.
C<char>            Change   replace next character by <char>.
D                  Delete   delete next character.
E                  Exit     end editing and reprocess text.
F<char>            Find     move pointer to next copy of <char>.
I<string><CTRL D or ESCAPE>
                   Insert   insert <string> after pointer.
K<char>            Kill     delete all chars until <char>.
P                  Print    print statement from pointer to end.
Q                  Quit     give up.
```

Making corrections 301

```
S<string><CTRL D or ESCAPE>
            Search   search for first copy of <string>,
                     leave pointer just before it.
<space>     ---      move pointer right one place.
?           (Help)   display this list.
```

A note on multi-line statements: the break between lines is treated as one character, corresponding to the RETURN key. If the imaginary cursor is just before the last character of a line, hitting the space bar puts it immediately after the last character, before the RETURN. If you now hit the space bar once, the cursor moves to the beginning of the next line. Or you can delete the RETURN by using D, or change it to a blank by using C, thus joining the two lines. You can break a line at any point by Inserting a RETURN: hit I, hit RETURN, hit ESCAPE.

6.3.3. Editing earlier statements

We have just explained how the last statement can be edited in some implementations of REDUCE 3.2. You can also edit (and re-execute) any earlier statement from the current REDUCE session, provided you know the prompt number it was assigned. (We are speaking here of lines typed by hand, not statements read in from a file.) For example, if you had at some time in the session entered the line

    23: AB5 := 7 * AB6;

(the "23:" being the prompt produced by REDUCE), and want now to enter

    AB25 := 7 * AB26;

without actually typing it, you can enter the command **ED 23$**. This retrieves and reprints line 23 for you, and allows you to edit it -- in this case, by inserting a "2" in two places. When you eventually type the E edit command, and strike RETURN, the modified statement is processed.

Of course if you change your mind (for example, if the line you wanted to modify and re-process turned out not to be line 23 after all), the Q (quit) edit command is available.

If you don't know the number of the line you need to edit, but know it's one of the -- say -- last 10, the command DISPLAY 10$ can be used in order to locate it. As the reader recalls from the Overview chapter, this command lists the 10 most recent commands (in reverse order). DISPLAY ALL$, or DISPLAY 99999$ (any huge number), will list all commands back to the beginning of the session. This produces an unreasonable amount of output if the session is well under way. If you have an approximate idea of where the desired statement is, a useful alternative is to hunt for it using ED n$ followed by Q (for QUIT), with trial values of n. (You will recall that ED n$ prints out line n, since it thinks you want to edit it.)

6.3.4. Correcting syntax errors in files

Suppose that there is a syntax error -- such as a missing parenthesis or extra parenthesis -- in a line of a file being read by an IN command. The IBM-PC implementation of REDUCE only gives you a choice of continuing, ignoring the defective statement, or (at least temporarily) aborting the reading from the file. (In the latter case you might then type in, manually, the correct version of the statement, and then enter CONT$.)

The DEC 20 version of REDUCE prints the defective statement (or what it interprets to be the defective statement), and asks you EDIT?. If you answer "Y", the standard text editor of the system is called in, and is automatically set to edit the file, starting with the line on which the defective statement begins. When you've made the correction, and exit from the text editor, reading of the IN file is resumed, starting with the line where the defective statement began.

This DEC 20 feature doesn't work if the file was stored without sequence numbers.

6.3.5. Correcting files

After reading in a file, with IN, and using some of the procedures or data thereby brought in, the user may realize that some changes to the file would be desirable. The most elementary way to accomplish this is to end the REDUCE run, call in and use the text editor to correct the file, and start the REDUCE run over again.

If much time and effort has been invested in the REDUCE run already, the user may be reluctant to start over. In some implementations, such as the one available at this time for the IBM-PC, there is no alternative. But some systems allow you to suspend the REDUCE run, saving its entire current status, including the values of all variables and the definitions of all procedures, by exiting from REDUCE (perhaps by typing BYE$ or by hitting CTRL C), and typing a system command that might be called PUSH or FORK. Then you can use the text editor to make the corrections. Next, get back to the "saved" REDUCE by typing a command that might be POP, or POP followed by CONTINUE, or EXIT. Then, back in REDUCE, we can "IN" the corrected file.

The DEC 20 implementation has automated this entire process. On this system, typing EDIT "ABCD.RED"$ will suspend REDUCE, and set up the standard text editor to edit the named file, "ABCD.RED". When you finish editing, and exit from the editor in the usual way, you automatically resume the suspended REDUCE, and can "IN" the file if desired.

(This is actually a refinement of a more general feature of the DEC 20 version of REDUCE. PUSH$ is available as a REDUCE command in this system. When executed it saves the REDUCE session status and makes it possible to perform other tasks without disturbing the session status, even to call in a fresh copy of REDUCE to do some subsidiary calculation!)

6.3.6. Correcting procedures

Suppose PROCEDURE XYZ had been defined earlier in the REDUCE session, and needs modification. There are several ways this can be done.

If XYZ was typed in by hand, and you remember the statement number that appeared in the prompt at the time, the ED n$ just described can be used.

If XYZ was typed in by hand, and you don't remember the statement number (and can't or don't want to find it using DISPLAY ALL$), in some REDUCE implementations you can enter EDIT XYZ$. (For the IBM-PC, use EDITDEF, not EDIT.) REDUCE takes the definition in the form in which it has been stored -- in LISP -- and converts it back into a REDUCE form. This is not likely to be identical to the form in which you typed it, and, frequently, includes operations and notations with which you may not be familiar, but it should be decipherable. The REDUCE string editor is then brought into play automatically, and you can make the desired corrections. Upon typing E (for EXIT) the modified form of the procedure is stored.

If XYZ was read from a file by an IN command, on the IBM-PC EDITDEF can be used as just described. But you must realize that while the procedure XYZ stored within REDUCE will be changed, the original copy on the file from which it came is not. On the DEC 20, EDIT XYZ$ will take care of this. This version of REDUCE remembers the name of the file from which XYZ came. It will activate the system text editor, and position it at the first line of the procedure, to enable editing that portion of the file. When you finish editing, and leave the text editor, the file is modified, and the revised form of the procedure is read in from the file. (Note the difference between this, and using EDIT filename$. In the latter, the changed file is not automatically read in, and an IN filename$ is required.)

This use of EDIT, on the DEC 20, will not work if the file is stored without line numbers, or if the line numbers in the file have

Making corrections   305

been changed (e.g., the file resequenced) since the last time the file was read in.

Caution: If there is a file XYZ as well as a procedure XYZ, on the DEC 20 the EDIT XYZ$ edits the file, not the procedure. To preserve the automatic reading-in of corrected procedures, one should avoid the otherwise popular practise of storing procedures in files of the same name.

6.4. INPUT, RETRY, CMD

We've already mentioned, in the Overview chapter, that (at least in REDUCE implementations in which the lines entered are numbered) any earlier line, say line 23, can be re-entered by using the command INPUT 23; or INPUT 23$. By "line", here, is meant statement or expression or procedure definition or command: this can be more than one physical line if it doesn't fit on one line, or -- less frequently -- a fraction of a physical line, if several statements are combined on one line.

In the IBM-PC and DEC 20 implementations almost the same effect can be obtained by using the command ED 23$ and then typing E (for Exit). One difference is that INPUT 23$ reports "not found" if there was a syntax error in that input; ED 23$ will find the line, and of course allow editing to remove the syntax error (and/or make any other change). INPUT n and ED n refer to two different history lists to make their responses.

With INPUT the choice of terminator -- semicolon or dollar sign -- overrides the terminator at the end of the line being brought back. So with INPUT 23; the result is printed whether or not line 23 ended with ; or $, while with INPUT 23$ the result is not automatically printed. Line 23 itself is not printed out in any case.

With ED, it doesn't matter which terminator is used. The original line is always printed out. The terminator at the end of the original line -- more exactly, at the end of the line after any changes the string editor may have made in it -- determines whether the result of that line is automatically printed or not.

RETRY$** is just like **INPUT n$, but refers automatically to the most recent line that triggered an execution error message such as `Declare ... operator?` or `Can't find file` or `ZERO DENOMINATOR`. Of course there is no point to issuing a **RETRY$** unless the cause of the error has been removed meanwhile.

In the DEC 20 REDUCE implementation the command CMD is available, that does for input from files the same thing INPUT does for input from the keyboard. If ABCD.RED is a file intended for input using IN, and is saved with line numbers, entering the command **CMD "ABCD.RED", 1100$** causes the statement on (or beginning on) line 1100 of the file to be read in and obeyed. For the file ABCD.RED listed on page 293, **CMD "ABCD.RED", 300$** would cause the definition of FACT (lines 300-1000) to be read in, and **CMD "ABCD.RED", 1100$** would cause 2**FACT 3 to be calculated.

6.5. OUT, FORT, LINELENGTH

6.5.1. OUT to files

Output can be diverted from the terminal to a file by use of the command OUT followed by the file name in double quotes. Output goes to that file from then on, until another OUT changes the output file, or SHUT (described below) closes it. Output can go to only one file at a time, although many files can be open at once. When it goes to a file, the only outputs that go to the terminal are error messages if any.

If the file has previously been used for output during the current job, and not SHUT, the new output is appended to the end of the file. Any existing file with that name is deleted before the first use of that file name for output in a job, or if had been SHUT before the new OUT.

To shift output back to the terminal without closing the output

OUT, FORT, LINELENGTH

file, so that more can be added to the same output file later, the command **OUT T** (T for terminal) is used.

```
OUT "OFILE.TXT"$          % output from now on goes
                          % to the file OFILE.TXT,
.....                     % not to the terminal

OUT T$                    % output to OFILE.TXT suspended,
                          % output to terminal resumes
.....

OUT "OFILE.TXT"$          % output to OFILE.TXT resumes

.....

OUT "ABCD.EEE"$           % output switches to ABCD.EEE,
                          % OFILE.TXT remains open
```

The SHUT command takes the name of a file that has been used for output, and on which no more is to be written, and closes it. What this means is that the file is released to the operating system for storage on disk, and is no longer available for additional output. In most operating systems if the REDUCE run is terminated without SHUTing a file that has had output directed to it, the file is lost. If a file is shut and a further OUT command issued for the same file, the file is in effect totally erased before the new output is written.

```
SHUT "OFILE.TXT"$

SHUT "ABCD.EEE"$
```

If it is the current output file that is shut, output will switch to the terminal. No **OUT T$** is necessary in this case.

The output sent to a file will normally be in the same form that it would have on the terminal. In particular, $X**2$ would appear on two lines, with an X on the lower line and a properly positioned 2 on the line above. If the purpose of the file is for later listing for human consumption, this is fine. But if the output file is intended for saving results to be read in later, using IN, this is not an appropriate form, and we must enter the command **OFF NAT$** to suppress the raising of the exponents.

OFF NAT$ also has another effect: each expression that is output will end with a dollar sign terminator to make use for input possible.

Suppose, for example, that we want to save the value of XYZ on the file OFILE.RED for later re-entry using IN.

```
OFF NAT$

OUT "OFILE.RED"$

XYZ:=XYZ;         % will output "XYZ := " followed by
                  % the value of XYZ, then a "$"

SHUT "OFILE.RED"$    % save OFILE.RED,
                     %    return to terminal output

ON NAT$           % restore usual output form
```

If you use OUT to create an OFF NAT file, and to examine it by typing it out or using a text editor, you may notice lines ending with some unexpected code such as "^z". Normally, IN doesn't accept numbers or variable names that are broken between lines. This code is a means for hyphenating the output, and is so recognized by IN, in some implementations of REDUCE.

6.5.2. FORT

Still another variant of file output is available if the output is to be embedded in the source file of a FORTRAN program. This may be desirable if the purpose for constructing an expression or collection of expressions was to evaluate them for many sets of numerical values. FORTRAN is much faster for this than is REDUCE.

The command ON FORT$ causes the output to be FORTRAN-compatible: columns 1-6 are left blank (except for a column 6 dot on continuation lines); exponents are written with ** "in line"; coefficients (but not exponents) have decimal points; "=" signs are used in place of ":=" ; and so on.

OUT, FORT, LINELENGTH 309

```
ON FORT$

OUT "FFILE.FOR"$

B := (X+Y)**10/(U+V)**2;

123*B;

SHUT "FFILE.FOR"$

OFF FORT$                    % return to normal
```

Let us see the generated file FFILE.FOR:

```
B=(X**10+10.*X**9*Y+45.*X**8*Y**2+120.*X**7*Y**3+
. 210.*X**6*Y**4+252.*X**5*Y**5+210.*X**4*Y**6+120.
. *X**3*Y**7+45.*X**2*Y**8+10.*X*Y**9+Y**10)/(U**2
. +2.*U*V+V**2)

ANS=(123.*(X**10+10.*X**9*Y+45.*X**8*Y**2+120.*X
. **7*Y**3+210.*X**6*Y**4+252.*X**5*Y**5+210.*X**4
. *Y**6+120.*X**3*Y**7+45.*X**2*Y**8+10.*X*Y**9+Y
. **10))/(U**2+2.*U*V+V**2)
```

Notice that the output generated by the second expression, which did not include an assignment, was automatically directed to a FORTRAN variable named ANS. If a different default name is desired, the command VARNAME is used before writing to the file:

```
VARNAME XXX$             % changes the default to "XXX ="
```

The ON FORT output mode is willing to make up to 20 continuation lines. This limit is controlled by the value of the variable CARDNO!* which the user can change if his version of FORTRAN allows fewer or more continuation lines. If the expression is too long to fit in that many lines, REDUCE breaks it up into parts called ANS1, ANS2, and so on, that are small enough to fit, and then combines the values of the parts. (After VARNAME XXX$ these parts are called XXX1, XXX2, and so on.)

It is assumed that the formulas put on the file will be used for "real" (i.e., "floating point") calculations, so all the coefficients (but not the exponents) are followed by dots. If the assumption is false, and integer arithmetic is wanted, the command OFF PERIOD$ can be used to prevent the creation of these dots.

6.5.3. LINELENGTH and FORTWIDTH!*

REDUCE assumes that the terminal can handle lines of 69 characters. This is also the maximum length line it will write on a file. To change this, say to 120, use the command

LINELENGTH 120$

The line length can of course be made shorter, too.

To find out the current line length, call LINELENGTH as a function of no arguments:

LINELENGTH ();

120

The LINELENGTH is ignored in ON FORT mode. However, there is a (slightly different) method to specify line length for FORTRAN-style output. FORTRAN-style output always starts in column 7 (8 for continuation lines) and normally goes to approximately column 64. This "64", however, can be changed: it is 6 less than the value of the variable FORTWIDTH!* which is initally 70. (The "6 less" is a margin of safety in the output routines to accomodate situations that sometimes occur at the end of the line.) The FFILE.FOR example just given was produced with

FORTWIDTH!* := 61$

to make it fit the pages of this book better!

6.6. COMPILE, FASLOUT, LOAD

6.6.1. Fast-loading files/fast-running procedures

In many implementations of REDUCE it is possible for the user to take any set of REDUCE commands, such as those one might have on a file for IN, and build a faster-loading version of them. Procedure definitions included among the commands are stored in a fashion that also execute much faster -- as much as ten times faster. (We refer to them as being "compiled".)

To create a fast-loading file to be referred to as ABCD, enter the command

    FASLOUT "ABCD"$

Now, type the commands, or use IN to read them from one or more files. (If you use a semicolon after the IN command, the contents of the file being translated is also printed out, so use with caution if the file is lengthy! Normally, a dollar sign should be used for the IN command to suppress the printing.) When all the input destined for storage in fast-loading form has been entered, type

    FASLEND$

In some implementations, the file resulting ABCD is automatically given the extension FAP, so its full name is ABCD.FAP. So we call such files "FAP" files. In some versions, the file is automatically stored in a special subdirectory, perhaps called FASL or RFASL.

To load such a file, use

    LOAD ABCD;   or   LOAD "ABCD";

depending on the implementation. Don't include ".FAP". (In some versions of REDUCE, LOAD is called FLOAD, which is short for "fast load".)

Even implementations of REDUCE that don't give the user the opportunity to create fast-loading files may support fast-loading files created by the system programmers, that can be loaded if needed. In some implementations this loading must be done by the user explicitly; in others, it happens automatically if one of the principal procedures in that file is needed, and the file has not been loaded before. The procedures for INT, FACTOR mode (and FACTORIZE), SOLVE, and PART, with their support procedures, are likely candidates for such handling. So, for that matter, are the procedures related to FASLOUT for <u>creating</u> fast-loading files.

A point to keep in mind is that once the procedures in a FAP file have been loaded in during a REDUCE session, either by using LOAD directly or through the loading-when-needed machinery, they remain in the user's work area, taking up space that may reduce the size of computation that can be done, or the speed with which smaller computations can be performed (on account of more frequent garbage collections).

Note that entering commands, definitions, etc. in order to create a FAP file doesn't also store them in the REDUCE environment. If after creating the FAP file you want to <u>use</u> the information in it, a LOAD is necessary.

One difference between IN and LOAD that should be noted is that while an ordinary ("slow-load") file can be read (using IN) several times during a REDUCE session -- for example, to pass the same series of commands over different data values, or to reset values or definitions that have been changed -- a second LOAD command for the same FAP file is ignored in most implementations. A given FAP file can be loaded only once in a REDUCE session. This is deliberate: this way, a procedure requiring something that is stored in a certain FAP file can LOAD it freely. The LOAD is ignored after the first time the procedure is executed, so no "... **REDEFINED**" messages are produced.

One more note: In some implementations no LOAD commands are allowed among the commands stored in a FAP file. An endless loop results upon trying to LOAD such a file.

It is not necessary for procedures to have been read from a FAP

COMPILE, FASLOUT, LOAD

file to be in compiled, therefore fast-running, form. If the command ON COMP$ has been given, any procedure whose definition is typed in or read in by IN is automatically compiled, in implementations that support this feature. One has to determine by experiment if the extra time taken for translating a procedure to compiled form is made up by the faster execution of the procedure.

The knowledgable and curious programmer who would like an idea of what compiling actually consists of can get it by entering the commands ON PLAP, PGWD$ after entering ON COMP$. In PLAP mode the "machine-independent macros" into which the procedures are first translated are printed out. In PGWD mode the actual assembly language instructions, and their machine-language equivalents, are displayed. (PGWD works only on some systems.)

6.7. TIME, SHOWTIME

In some REDUCE systems, the command SHOWTIME$ prints out the amount of time the computer was actually working for the user since the last SHOWTIME request or since the start of the REDUCE run. On mainframe computers that are being shared by several users in timeshared mode, this will be a fraction of the elapsed wall-clock time. The time is the sum of the times required to analyse the input, perform the required calculation, and format and print the output. (Frequently, the formatting time is the largest part of the total time!) Times are printed in milliseconds.

On some microcomputers the times reported can be very inaccurate, so the information should be used with caution.

The command ON TIME$ puts REDUCE into an automatic SHOWTIME mode. The time taken by every command is printed. Of course OFF TIME$ terminates this mode.

If the same REDUCE operation is done repeatedly, it is found that the required time is not constant. For example, on one system the time usually gets gradually less with each repetition, and then fluctuates slightly. Since this sort of thing doesn't happen in conventional computing environments, we digress to explain this phenomenon. The initial speedup occurs as the underlying LISP system reorganizes its symbol table ("OBLIST") to bring the symbols used in carrying out this operation to more accessible locations. Once this is completed the time fluctuates because of variations in the time required to build the necessary list structures.

After the time appears nearly stabilized there are occasional wild increases, of the order of 1000 milliseconds. These occur when LISP finds it necessary to reorganize its working area and carry out a so-called "garbage collection" pass. The greater the amount of memory with which REDUCE is working, the longer this takes but the less frequently is it necessary.

On some REDUCE systems the command ON GCGAG$ instructs REDUCE to print an informative message every time garbage collection takes place.

6.8. DEFINE

Suppose we find ourselves with an awkwardly-named variable, say VXQJMZB, and discover that we mistype it more often than we type it correctly. We can attempt to use LET to provide a simpler name as a synonym:

```
LET VV = VXQJMZB$
```

Then if we wanted, say, to set A equal to 2 times the value of VXQJMZB, we need only enter

```
A := 2 * VV$
```

But suppose we wanted to give VXQJMZB the value 123. The input

```
VV := 123$
```

would not accomplish what was wanted. (It would change the value of VV from VXQJMZB to 123, and not affect VXQJMZB at all.)

What is needed, instead of the LET, is

DEFINE   315

```
DEFINE VV = VXQJMZB$
```

Then any place you type VV (except within quotes) it is as if you had typed VXQJMZB: for example,

```
VV := 123$
```

would set VXQJMZB to 123.

DEFINE can also be used to provide new and easier names for REDUCE commands and operations:

```
DEFINE WR = WRITE$        % Abbreviation for WRITE

WR A," ",2**10$           % Try it

      A 1024              % It worked
```

Caution: once you define a symbol to be an abbreviation for another, you can't change the definition for the duration of that REDUCE run. To illustrate this, suppose later in the run we decide to use WR as an abbreviation for SQRT instead (unlikely as that may be). Let us see what happens.

```
DEFINE WR = SQRT$

WR 36;

          6               % Looks OK

WRITE 49;

          7               % ?????
```

The explanation is that once WR is DEFINEd to stand for WRITE, REDUCE treats it as if WRITE had been typed no matter where it appears -- even in another DEFINE. So in particular the command DEFINE WR = SQRT$ is treated as if it were DEFINE WRITE = SQRT$. From then on, "WRITE" doesn't mean WRITE!

DEFINE can also be used to provide abbreviations for complete expressions, not just single names.

DEFINE QUAD = A*X**2 + B*X + C;

Then any place you type QUAD (except within quotes) it is as if you had typed A*X**2 + B*X + C. This use of DEFINE does not provide any benefits over using LET because such expressions can not normally appear on the left-hand side of a ":=" sign. (Actually there is an esoteric use of expressions on the left-hand side of ":=" signs: see page 104.)

We can use DEFINE to provide new names for single names, and for complete expressions. We can not use it for abbreviating partial expressions. Suppose we often need the square root of the numerator of various expressions, as in SQRT NUM A; or SQRT NUM (B+C);. "SQRT NUM" is a part of these expressions. Try to abbreviate it:

DEFINE SN = SQRT NUM;

SN (25/7);

***** SQRT UNDEFINED FUNCTION - EVAL

This or some other cryptic error message, arising from the underlying LISP system, is not the number 5 we expected. We can't undertake to explain why precisely this was the generated message, but observe that it is exactly the same message that we'd get if SN were replaced not by SQRT NUM but by (SQRT NUM):

(SQRT NUM) (25/7);

***** SQRT UNDEFINED FUNCTION - EVAL

Of course the proper way to introduce SN would be as a procedure or a LET statement:

PROCEDURE SN X$ SQRT NUM X$

SN (25/7);

DEFINE   317

6.9. Tracing

A facility is available in most versions of REDUCE for tracing procedure calls. If AAA is a procedure, the command TR AAA$ changes AAA so there is a printout of the procedure name AAA and the values of its arguments (if any) every time AAA is called, and a printout of the result when it finishes.

The output of the argument values and of the result have to be examined in a spirit of tolerance, because they are not in algebraic but in LISP form. Indeed, the arguments are in the somewhat obscure so-called "prefix" form while the result may be in prefix form or in the much more obscure SQ (Standard Quotient) form. However, often all that needs to be known are the facts that the procedure was called at all, and that it completed its job.

A single TR command can put several procedures into trace mode:

TR AAA,BBB,CCC$

(AAA BBB CCC)

As indicated, when TR is called it prints out a list of the procedures succesfully modified for tracing.

The tracing of one or more procedures may be turned off by using the command UNTR.

UNTR AAA,CCC$

(AAA CCC)

The tracing of a procedure is also turned off if it's redefined by entering a new definition, or even reentering the original definition.

An additional kind of trace that is available in some implementations of REDUCE reports the results of := assignments during the execution of a procedure. The command for starting this kind of tracing is TRST AAA$. It works only if TR AAA$ is also in effect. The procedures that can be traced in this fashion must not be in compiled form, and must have the procedure body enclosed in a BEGIN ... END. Only assignment performed directly inside this BEGIN ... END (that is, not within grouping symbols << >> or constructs like FOR ... DO) are reported.

An UNTRST AAA command only removes the assignment tracing mode from AAA. To remove both the assignment tracing and the argument-result tracing, UNTR AAA must be used. Assignment tracing without argument-result tracing is not possible.

6.10. Expression input

The reader who is familiar with programming languages such as Basic, FORTRAN, Pascal, and the like, may suddenly realize that we have not mentioned anything corresponding to the command variously called READ, INPUT, or ACCEPT that these languages have. That is, we haven't explained how to have a REDUCE "program" call for the user to supply, by typing in, a value (a number or an algebraic expression) that the "program" needs to do its job. We have relied, instead, on typing in complete assignments like X := 123$ or A := P**2 + Q**2$ among the statements of which our input to a REDUCE session is composed. Actually there <u>does</u> exist a procedure, XREAD, that enables one to type in expressions. The way to use it is not obvious, however, so we shall present it in stages, somewhat in the style of our Case Studies chapter.

If we have to store certain expressions in array places A(1) through A(5), we <u>could</u> enter

A 1 := ...$

A 2 := ...$

. . . .

A 5 := ...$

but we are likely to get tired of repeatedly typing the "A", the place number, and the ":=" sign on each line.

Expression input

Suppose we first typed in this FOR statement:

```
FOR I := 1:5 DO A(I):=XREAD()$
```

as, say, line 25 of the session. As execution of the FOR statement begins, the built-in procedure XREAD, which has no parameters and so has to be called as XREAD(), with empty parentheses, will be activated. It prints out the current prompt symbol, here "25:", and suspends progress through the FOR loop until the user types an expression, followed by a semicolon or dollar sign terminator and a RETURN. As soon as the RETURN is struck, the value of the XREAD() will become the value of the expression typed, and will be stored in A(1), because I is now equal to 1 and so A(1):=XREAD() is executed.

Advancing in the FOR loop, I will become 2, and XREAD will again prompt for input. The prompt symbol remains "25:", because REDUCE is still executing an aspect of statement 25. After a terminator is typed and RETURN is struck, the new expression typed will be stored in A(2).

After five such inputs, A(1) through A(5) will have received their values, and REDUCE issues the prompt "26:" to show that it is ready for the next statement. The entire transaction may look like this:

```
24: ARRAY A 5$

25: FOR I := 1:5 DO A(I):=XREAD()$
25: AAA;
25: BBB;
25: CCC;
25: DDD;
25: EEE;

26:
```

Can this FOR statement be in a file? No, not as it stands, for direct use. If it were being read in from a file, and executed immediately, XREAD would expect to read the expressions from the same file, not from the user's terminal! Input into REDUCE can come from only one place at a time -- either the terminal, or some one file. But what we can do is to package the FOR loop as a procedure, read in that procedure from the file, and call that procedure from the terminal once the file has been completely read in.

The procedure we are defining needs no parameters, so it is more convenient to use the alternative notation in which it is represented as a LET statement instead of a formal procedure. Thus we would have, somewhere in file ABCD.RED:

```
ARRAY A 10$

LET R1 =
    FOR I := 1:5 DO A(I):=XREAD()$
```

Reading in the file and executing R1:

```
37: IN "ABCD.RED"$

38: R1$
38: AAA;
38: BBB;
38: CCC;
38: DDD;
38: EEE;

0

39:
```

(We'll discuss, later, where that zero came from.)

It would be nice to have a message print out reminding the user what manner of input is desired. In addition, those repeated numerical prompts should be changed. We should adopt some special prompt symbol, say ">>", to distinguish this call for input from others. A procedure, SETPCHAR, for temporarily changing the "prompt characters", is built into RLISP but is not normally available in REDUCE proper. Fortunately we can remedy this restriction by writing LISP in front of the procedure name. The LISP SETPCHAR must be executed before each XREAD call, as shown in version R2 of our solution to the problem of input into an array:

Expression input

Somewhere in file ABCD.RED:

```
LET R2 =
<<WRITE "ENTER 5 VALUES FOR ARRAY A"$
  FOR I := 1:5 DO
        <<LISP SETPCHAR ">>";
          A(I):=XREAD()>>
>>$
```

Reading in the file and executing R2:

```
47: R2$

ENTER 5 VALUES FOR ARRAY A
>>AAA;
>>BBB;
>>CCC;
>>DDD;
>>EEE;

0
```

Why are we getting a zero printed at the bottom? We would expect this if we had used R2; as our call, but we used R2$ so no meaningless "result" should print out. The explanation is that when REDUCE has completed the R2 action, it checks the most recently typed terminator to determine whether or not to print a "result". In this case, that terminator was not the dollar sign following the R2, but the semicolon after the last expression entered, in our example EEE;.

If that zero really bothers us, we could make a practice of ending each expression with a dollar sign instead, but on most terminals that is somewhat less convenient. We could add another "LISP" line into our program to call an RLISP procedure to make REDUCE think that the terminator was a dollar sign even if it wasn't, but that goes against the philosophy guiding this book. (LISP SETPCHAR was presented because there is no alternative possible.) Now carefully compare the following R3 with the R2 we have just used:

```
LET R3 =
<<WRITE "ENTER 5 VALUES FOR ARRAY A"$
  FOR I := 1:5 DO
        <<LISP SETPCHAR ">>";
          A(I):=XREAD()>>;
""»>$
```

The only change is in the very last line. We've inserted a pair of quotation marks in front of the closing bracket. This, the "empty string", is the last expression in the main "<< >>" bracket, so it is taken as the "value" of R3. There are no characters in it. So when the "value" of R3 prints out, nothing visible prints out!

A final improvement would be to label the individual calls for input. This change is easy:

```
LET R4 =
<<WRITE "ENTER 5 VALUES FOR ARRAY A"$
  FOR I := 1:5 DO
        <<WRITE "A ",I," := ";
          LISP SETPCHAR ">>";
          A(I):=XREAD()>>;
""»>$

4: R4;

ENTER 5 VALUES FOR ARRAY A

A 1 :=
>>AAA;

A 2 :=
>>BBB;

A 3 :=
>>CCC;

A 4 :=
>>DDD;

A 5 :=
>>EEE;
```

Unfortunately there is no easy way, without going into RLISP, to have the prompt symbol >> print on the same line as the :=.

Expression input   323

A word of caution: Be especially careful to check the expression
you enter in response to an XREAD before you strike RETURN. If
what you type contains a syntax error (such as unmatched
parentheses), the entire FOR statement is aborted, and you have no
opportunity to correct the faulty item or to input the rest.

6.11. Lost in LISP?

We close this chapter on Running REDUCE with a brief
mention of an aspect of the REDUCE system that is intended for
expert programmers, but which occasionally intrudes into the world of
the ordinary REDUCE user.

Access to the RLISP language, in which most of REDUCE itself
is written, can be had by typing the command LISP$ or alternatively
SYMBOLIC$. The RLISP language is similar in overall appearance to
the REDUCE language we've been studying, but has as its basic
operations the manipulation of lists of symbols rather than of
algebraic expressions.

In RLISP if a variable has no assigned value, asking for its
value doesn't produce its name, as is the case for clear variables in
REDUCE, but an error message, "Unbound variable". ("Unbound"
means that no value has been "bound" to the variable. It has
nothing to do with "unbounded" meaning "infinite".) If you have
somehow stumbled into RLISP, as shown by a profusion of "Unbound
variable" messages, entering the command

   ALGEBRAIC$

should bring you safely back to the familiar algebra language you
have grown to love.

The RLISP language is, in turn, a superstructure on an
underlying LISP system. LISP has the same intended use as RLISP,
but the appearance of a LISP program is entirely different. Gone is
the := assignment symbol, and gone is the semicolon or dollar sign.
Instead, one sees deeply nested parentheses, and words like SETQ.

It is embarrasingly easy to wander into the LISP world from REDUCE. If you manually type in an END$ when REDUCE doesn't expect it -- because there is no BEGIN waiting for a matching END -- a message "ENTERING LISP" tells you that you have crossed the border. Once there, most lines you enter lead to "Unbound variable" messages, including ones of the forms

```
***** Unbound variable !;

***** Unbound variable !$

***** ABC!; Unbound variable
```

In some systems, the excursions to LISP-land are automatically short, because any LISP error message returns you to the REDUCE algebra system. On others, you must type the LISP command

(BEGIN)

(note the required parentheses, and the absence of a semicolon or dollar sign) to get back to REDUCE.

Note that this happens only if the extra END is entered from the keyboard. If it occurs in a file, the extra END$ just causes the reading of the file to be cut short, as we've already said.

Index

$ 18

% 18

EOF 295
** 11
^ 12

ALGEBRAIC 323
Algebraic numbers 275
Algebraic topology 286
ALLFAC 144, 211, 258
AND 39
ANS 309
ANS1 309
Anti-symmetric operator 96
ANTISYMMETRIC 97
ARBCOMPLEX 73
ARBINT 76
Array 45, 190
Array bound 46
Array name 198
Assigned value 14, 134
Assignment 12, 56
Assignment to expressions 104
Automatic copy 194

BACK SPACE 297
BEGIN ... END 108, 111, 176, 324
Bell 297
BIGFLOAT 23, 159, 233
Binomial coefficients 286
Boolean 37, 95, 114, 149
BYE 291

Call by value 196, 207

CALLLET 227, 261
Cancellation 24
Cardan's formula 74, 278
CARDNO!* 309
Change 215
Clear 13, 21, 48, 52, 54, 55, 56, 81, 87, 103, 171
CLEAR array place 49
Clearing formal parameters 198
CMD 306
COEFF 65, 81, 103
Coefficients 65
Comments 18
Common denominator 130
Common factors 24, 126
COMP 313
Comparison 38
Compiled procedures 311, 313
Complex conjugate 191
CONT 296
Controlled variable 32
Copy variable 194, 207
Copying arrays 47, 56
Copying matrices 56
CORE 30
COS 22
CTRL C 297
CTRL G 291, 297
CTRL U 297
CTRL X 297
CTRL Z 308

Decimal approximation 37
Default ordering 149, 153
DEFINE 315
DEG 66, 211

DELETE 297
DEMO 296
DEN 24, 212
Denominator 24
DEPEND 28, 29
Derivative 213
DF 26, 97, 103, 201, 213
Differentiation 26
Dimension 46, 47
DISPLAY 22, 302
DIV 141
Dollar sign 18
Domain modes 153
Double sum 36
Dummy statement 111

E 23, 32, 160
ECHO 294
ED 298
ED n 301, 305
EDIT file name 303
EDIT procedure name 304
EDITDEF 304
Editing files 303
Elapsed time 313
Ellipse 6
Empty statement 111
Empty string 322
END, extra, in file 295
END, extra, typed 324
End-comment convention 295
Endless loop 20
Environment 79
Error exit 204
Euclidean Algorithm 239
Evaluation 133
EXP 90, 124, 139
Exponents 14, 20
EXPT 280

FACT 194
FACTOR 145, 266
FACTOR command 136
FACTOR mode 71, 138, 154
Factorial 205, 207
FACTORIZE 68, 154
FALSE 37
Families of polynomials 252
FAP 311
FASLEND 311
FASLOUT 311
Fast-loading files 311
FLOAD 311
FLOAT 37, 154, 233
Floating point 154
FLUID 123
FOR ... DO 32, 319
FOR ... PRODUCT 36
FOR ... SUM 34, 51
FOR ALL X LET ... = ...
    88, 256, 280
FOR ALL X SUCH THAT ...
    LET ... = ... 94
Formal parameters 183, 194
FORT 149, 308, 310
FORTRAN 308
FORTWIDTH!* 310
Function 175, 184

Garbage collection 218, 312, 314
GCD 25, 26, 127, 131, 152
GCD function 238
GCGAG 314
General matrices 57
Global variables 188, 190, 202
GO TO 205
Greatest common divisor 127, 238

Grouping 33, 37, 107, 246, 269

Hearn, A. C. 9
HIPOW!* 65

I 23, 32, 107, 191
IF ... THEN ... ELSE statement 114
IF ... THEN statement 113
IN 59, 292
Indeterminate 23
INPUT n 22, 78, 193, 305
INT. 29, 154, 268
Integers 153
Integration 29
Interchange 43, 186
INTERVL 77

Kernel 23, 28, 219, 270
KORDER 84, 127, 136, 150, 153, 158, 164, 233
KORDER NIL 150

Label 205
LCM 131, 237
LCOF 230
Leading coefficient 230
Leading term 230
Least common multiple 130, 236
Legendre polynomials 186, 190
LET 77, 171, 314
LET DF(...) = 97
LET operator = ... 85
LET power = ... 79
LET procedures 206
LET product = ... 82
LET sum = 151

LET sum = ... 83
Lexical scope 203
LINELENGTH 310
Lines 19
LISP 13, 320, 323
LISP command 323
LIST 145
LOAD 29, 68, 71, 139, 311
Local variables 33, 183, 188, 189
LOG 22
LOWPOW!* 66
LST 72
LTERM 230

Machine language 313
Machine-independent macros 313
Maclaurin expansion 44, 251
Main operator 118, 122
MAINVAR 215
Marti, J. 10
MAT 58
MATCH 82, 263
Matrices 54
MATRIX 54, 190
MCD 88, 132, 142
Minus 101
Mode name errors 123
Modes 123
MODULAR 165, 214
Modular arithmetic 165
MSG 225
Multi-dimensional array 47, 67
MULTIPLICITY 72

NAT 148, 307
NERO 60, 147
Newton polynomials 260

NIL   111, 183, 190
NODEPEND   28
NONCOM   282
NUM   24, 212
NUMBERP   39, 95, 209
Numbers   12
Numerator   24
NUMVAL   23, 159

OFF   123
Omitted parentheses   48, 53, 182
ON   123
Operator   45, 50, 190, 253
Options   123
OR   38, 39
ORDER   134, 153, 212
ORDER NIL   135, 211
ORDFP   150
ORDP   149
Orthogonal functions   265
OUT   148, 306
OUT T   307

Parameter   182, 185
Parameterless procedure   172
PART   117
Partial fractions   26
Partial functions   208
PAUSE   296
Percent sign   18
PGWD   313
PI   23, 160
Pi (product symbol)   36
Place   46, 50
PLAP   313
Position   46
Power symbol **   11
Powers   65

PRECISION   159
Prefix form   317
PRI   149, 279
Procedure   150, 169
Prompt   22, 320
Pseudoremainder   234
PUSH   303

Quaternions   282, 283
QUIT   291
Quotes   105, 213

Random numbers   219
RAT   143
RATIONAL   163, 218, 234
Rationalizing denominators   191, 268
READ   318
Real   154
Recursion   193
RED   292
REDERR   204
REDUCE.INI   295
REDUCT   230
REMAINDER   168, 232
REMFAC   137
REPEAT ... UNTIL   40
Reserved words   13
RESUBS   133, 245
Result   18, 175
RETRY   52, 193, 306
RETURN   111, 174, 176
RETURN WS :=   177
Returning multiple values   178
RLISP   8, 169, 320, 322
Roundoff   154
Rules   78

SAVEAS   22, 93

SCALAR 189
Semicolon 18, 40
Series 36
SETMOD 165
SETPCHAR 320
Setting a part 120
SHOWTIME 313
SHUT 307
Sigma (summation symbol) 34
SIN 22, 160
SOLVE 71, 154
SOLVEINTERVAL 77
SOLVEWRITE 72
SQRT 22, 160
SQRT, sign of 23
Standard functions 160
Standard quotient form 317
Steenrod Squares 286
STEP ... UNTIL 32
String editor 298
SUB 42, 53, 199, 268, 270
Subroutine 175
Substitution 42
Substitution environment 79
Substitution rules 78
Subtraction 101
SUM 13, 173
Superparentheses 112, 158, 269, 286
Surds 278
Switches 123
SYMBOLIC 323
SYMMETRIC 96
Symmetric operator 95
Syntax errors 302

T 183, 190, 253
Terminator 18
TIME 313

Top level 190
TR 317
Tracing 31, 70, 141, 317
TRFAC 70, 141
Trigonometric identities 7, 22
TRINT 31
TRST 317
TRUE 37
Tschebycheff polynomials 252

Unary + and - 17
Unbound 323
UNTR 317
UNTRSRT 318

Value 12
Variables 11, 44
VARNAME 309
VECTOR 56

WEIGHT 102, 198
WHILE ... DO 36
Workspace 21
Wrong answers 23, 177, 282
WS 21, 92, 175
WS n 22
WTLEVEL 102

XREAD 318

OHIO UNIVERSITY LIBRARY

Please return this book as soon as you have finished with it. In order to avoid a fine it must be returned by the latest date stamped below.

SEP 11 1988

SEP 24 1988

QUARTER LOAN
JAN 3 1990

NOV 8 1989

FACULTY LOAN
JUN 15 1996

APR 01 1996

JUN 30 1988